AF614703

Methods in Molecular Biology™

Series Editor
John M. Walker
School of Life Sciences
University of Hertfordshire
Hatfield, Hertfordshire, AL10 9AB, UK

For other titles published in this series, go to
www.springer.com/series/7651

MicroRNAs in Development

Methods and Protocols

Edited by

Tamas Dalmay

School of Biological Sciences, University of East Anglia, Norwich, UK

Editor
Tamas Dalmay, Ph.D.
School of Biological Sciences
University of East Anglia
Norwich, UK
t.dalmay@uea.ac.uk

ISSN 1064-3745 e-ISSN 1940-6029
ISBN 978-1-61779-082-9 e-ISBN 978-1-61779-083-6
DOI 10.1007/978-1-61779-083-6
Springer New York Dordrecht Heidelberg London

Library of Congress Control Number: 2011923430

Printed on acid-free paper

Humana Press is part of Springer Science+Business Media (www.springer.com)

Preface

How do a lion or an orchid develop from a single cell? Answering this question in detail has fascinated developmental biologists for a long time. Plants and animals can have simple or very complex body organization but they all derive from a single cell, the fertilised egg. This cell divides and the progeny cells divide many–many times to build the entire body but the genetic information does not change during these cell divisions. Therefore, all our cells contain the same genetic information but there are many different tissues with specialised functions in our body. These tissues are different from each other because a different set of proteins are present in the cells that make up a certain tissue. The reason for this is that only a certain set of genes are active in each cell. Gene expression is a relatively complex process; therefore, it can be regulated at several layers. First the chromosomal DNA is transcribed into mRNA and this step is regulated by various mechanisms. For example, transcription factor proteins can activate or suppress the transcription and certain modifications of the DNA and the histon proteins that package the DNA also regulate transcription. The mRNAs are then processed and translocated to the cytoplasm where they are translated into proteins. Accumulation level of a protein can be regulated at the mRNA processing stage, mRNA stability level and through the half-life of the protein itself. One of the most recently recognised regulatory layers involves short RNAs to regulate the translation efficiency of mRNAs. These short RNAs are called microRNAs (miRNAs) since these molecules are very short, only 21–24 nucleotides.

The first miRNA (*lin-4*) was discovered in 1993, although it was not called a miRNA [1]. The next miRNA (*let-7*) was discovered 7 years later [2], but at that time these molecules were called small temporal RNAs because of their specific expression pattern during certain developmental transitions. The name miRNA was coined in the following year when three groups identified many short RNAs similar to *lin-4* and *let-7* in different organisms and because some of them were expressed all the time, the name "short temporal" was changed to miRNA [3–5]. miRNAs soon became one of the most intensively studied subjects in molecular biology. It is now clear that many mammalian genes are regulated by miRNAs, therefore understanding the role of miRNAs in development and disease is an important but difficult task.

One could say that working with miRNAs is not different from working with other RNAs. However, the very small size of miRNAs often requires specific techniques to study them and standard protocols (that are used for mRNA analysis) either cannot be used or important modifications have to be made. This book describes protocols for investigating miRNAs in plant and animal development. The chapters fall into three sections. Chapters 1–6 describe various techniques to detect and profile miRNA expression either spatially or at different time points. In situ hybridisation can establish where the miRNAs are expressed and northern blot, qPCR, deep sequencing, and array can be used to profile the expression of miRNAs at different developmental stages. Deep sequencing also has the potential to discover new miRNAs. Chapters 7–10 are protocols to manipulate the activity of miRNAs in various organisms. These approaches are very useful to learn more about the function of miRNAs in developmental processes. Finally Chapters 11–15 describe different methods to identify and validate miRNA targets in animals and plants.

Norwich, UK *Tamas Dalmay*

References

1. Lee RC, Feinbaum RL, Ambros V. (1993) The C. elegans heterochronic gene lin-4 encodes small RNAs with antisense complementarity to lin-14. *Cell*, **75**, 843–54.
2. Reinhart BJ, Slack FJ, Basson M, Pasquinelli AE, Bettinger JC, Rougvie AE, Horvitz HR, Ruvkun G. (2000) The 21-nucleotide let-7 RNA regulates developmental timing in Caenorhabditis elegans. *Nature*, **403**, 901–6.
3. Lagos-Quintana M, Rauhut R, Lendeckel W, Tuschl T. (2001) Identification of Novel Genes Coding for Small Expressed RNAs. *Science*, **294**, 853–858.
4. Lau NC, Lim LP, Weinstein EG, Bartel DP. (2001) An abundant class of tiny RNAs with probable regulatory roles in Caenorhabditis elegans. *Science*, **294**, 858–62.
5. Lee RC, Ambros V. (2001) An extensive class of small RNAs in Caenorhabditis elegans. *Science*, **294**, 862–4.

Contents

Contributors

DITTE ANDREASEN • *Exiqon A/S, Vedbaek, Denmark*
DANIEL BARAJAS • *Department of Environmental Biology, Centro de Investigaciones Biológicas – CSIC, Madrid, Spain*
MICHAELA BEITZINGER • *Laboratory for RNA Biology, Center for integrated protein science Munich (CIPSM), Max-Planck-Institute of Biochemistry, Martinsried, Germany*
YA-WEN CHEN • *Institute of Molecular and Cell Biology, 61 Biopolis Drive, Proteos, Singapore 138673, Singapore*
STEPHEN M. COHEN • *Institute of Molecular and Cell Biology, 61 Biopolis Drive, Proteos, Singapore 138673, Singapore*
REBECCA T. COOK • *Department of Biology, University of Pennsylvania, Philadelphia, PA, USA*
TAMAS DALMAY • *School of Biological Sciences, University of East Anglia, Norwich, UK*
MATTHEW W. ENDRES • *Department of Biology, University of Pennsylvania, Philadelphia, PA, USA*
JOSÉ MANUEL FRANCO-ZORRILLA • *Department of Plant Molecular Genetics, Genomics Unit, Centro Nacional de Biotecnología – CSIC, Madrid, Spain*
JOZSEF GAL • *Department of Molecular and Cellular Biochemistry, University of Kentucky, Lexington, KY, USA*
BRIAN D. GREGORY • *Department of Biology, Penn Genome Frontiers Institute, Graduate Group in Genomics and Computational Biology, University of Pennsylvania, Philadelphia, PA, USA*
ERICKA R. HAVECKER • *Department of Plant Sciences, University of Cambridge, Cambridge, UK*
ZOLTÁN HAVELDA • *Agricultural Biotechnology Center, Gödöllő, Hungary*
ERAN HORNSTEIN • *Department of Molecular Genetics, Weizmann Institute of Science, Rehovot, Israel*
NANA JACOBSEN • *Exiqon A/S, Vedbaek, Denmark*
SHARON KREDO-RUSSO • *Department of Molecular Genetics, Weizmann Institute of Science, Rehovot, Israel*
NATASHA KYPRIANOU • *Division of Urology, Department of Surgery, University of Kentucky, Lexington, KY, USA*
CÉSAR LLAVE • *Department of Environmental Biology, Centro de Investigaciones Biológicas – CSIC, Madrid, Spain*
SARA LÓPEZ-GOMOLLÓN • *School of Biological Sciences, University of East Anglia, Norwich, UK*

Pablo Andrés Manavella • *Max Planck Institute for Developmental Biology, Tübingen, Germany*
Gunter Meister • *Laboratory for RNA Biology, Center for integrated protein science Munich (CIPSM), Max-Planck-Institute of Biochemistry, Martinsried, Germany*
Peter Mouritzen • *Exiqon A/S, Vedbaek, Denmark*
Francisco E. Nicolas • *School of Biological Sciences, University of East Anglia, Norwich, UK*
Amy E. Pasquinelli • *Department of Biology, University of California, La Jolla, San Diego, CA, USA*
Ignacio Rubio-Somoza • *Max Planck Institute for Developmental Biology, Tübingen, Germany*
Roberto Solano • *Department of Plant Molecular Genetics, Genomics Unit, Centro Nacional de Biotecnología – CSIC, Madrid, Spain*
Dylan Sweetman • *School of Biosciences, University of Nottingham, Sutton Bonington Campus, Loughborough, UK*
Guiliang Tang • *Department of Plant and Soil Sciences & KTRDC, University of Kentucky, Lexington, KY, USA*
Xiaohu Tang • *Department of Plant and Soil Sciences & KTRDC, University of Kentucky, Lexington, KY, USA*
Xiaoqing Tang • *Department of Biological Sciences, Michigan Technological University, Houghton, MI, USA*
Éva Várallyay • *Agricultural Biotechnology Center, Gödöllöő, Hungary*
Ruifen Weng • *Institute of Molecular and Cell Biology, 61 Biopolis Drive, Proteos, Singapore 138673, Singapore*
Gene W. Yeo • *Department of Cellular and Molecular Medicine, University of California, La Jolla, San Diego, CA, USA*
Haining Zhu • *Department of Molecular and Cellular Biochemistry, University of Kentucky, Lexington, KY, USA*
Dimitrios G. Zisoulis • *Department of Biology, University of California, La Jolla, San Diego, CA, USA*

Chapter 1

In Situ Detection of microRNAs in Animals

Dylan Sweetman

Abstract

The detection of microRNAs (miRNAs) in situ presents several technical challenges. Although protocols for mRNA detection by in situ hybridization are well established, the small size of miRNAs makes their localization problematic. To overcome this, digoxygenin-labeled locked nucleic acid oligos have been used. These bind strongly and specifically to miR targets and make the identification of the precise spatiotemporal expression of these molecules possible.

Key words: LNA, In situ hybridization, microRNA detection

1. Introduction

To understand the biological roles of microRNAs (miRNAs), it is first necessary to know their expression patterns. Given that many miRNAs are predicted to regulate hundreds of potential targets, a detailed knowledge of the expression of both miR and putative target is required to establish any meaningful relationship between the two. There are two competing models for miR function: exclusive expression where miRNAs are expressed to prevent ectopic expression of their targets (1) and overlapping expression where miRNAs are present to modulate protein expression levels from their target mRNAs (2, 3). It has also been suggested that changes in miR expression patterns have contributed to evolution and that changes in miR expression patterns lead to morphological variation across species (4).

Whatever the function of miRNAs, the ability to determine their precise spatial and temporal expression patterns is critical to understand them. The development of locked nucleic acid (LNA) oligos has been a crucial step in the development of this technology,

Tamas Dalmay (ed.), *MicroRNAs in Development: Methods and Protocols*, Methods in Molecular Biology, vol. 732, DOI 10.1007/978-1-61779-083-6_1, © Springer Science+Business Media, LLC 2011

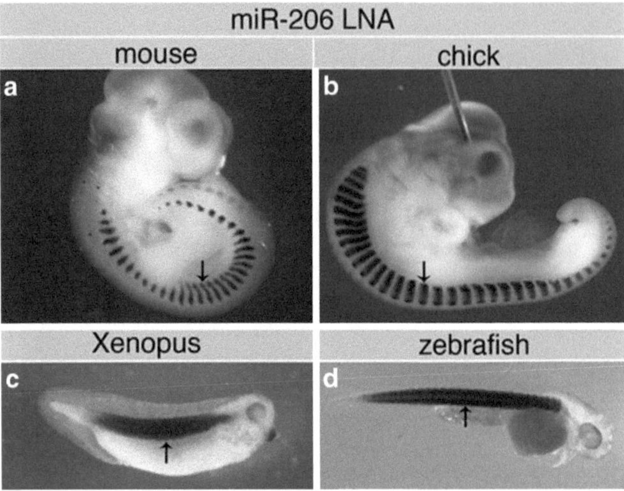

Fig. 1. Expression of miR-206 in mouse. (**a**), chicken (**b**), *Xenopus* (**c**), and zebrafish (**d**) embryos. Staining is seen in muscle precursor cells in somites in all these species (*arrows*). The same probe was used for all these embryos.

as they bind strongly to complementary sequences and also show remarkable specificity in detecting miR expression patterns (5, 6). Use of these modified nucleotides was crucial in the development of whole-mount in situ hybridization methods to detect miRNAs which were first used in zebrafish embryos (7), and subsequently embryos of other species such as chicken, mouse, and *Xenopus* (8, 9). In addition, it is also possible to combine LNA probes with normal antisense RNA probes for simultaneous detection of miRNAs and mRNAs (10) (see Note 11).

The method described here can be used to detect miRNAs in chicken, mouse, *Xenopus,* and zebrafish embryos. Figure 1 shows detection of the muscle specific miR-206 in the myotome (the embryonic source of muscle cells) in all these species using the same LNA probe. It will also distinguish between similar probes such as miR-1a and miR-206 which differ by only three base pairs. This is shown in Fig. 2. The specificity of these probes is apparent as miR-206 is detected only in the myotome, whereas miR-1 is detected in both myotome and heart muscle.

2. Materials

1. Diethyl pyrocarbonate (DEPC)-PBS – 1× PBS with DEPC added to 0.1%. Treat overnight in a fume hood and then autoclave before use (see Note 1).
2. 4% paraformaldehyde (PFA) in PBS – heat PFA powder in a fume hood in DEPC-PBS to dissolve. This can be frozen at –20°C in aliquots and thawed out prior to use.

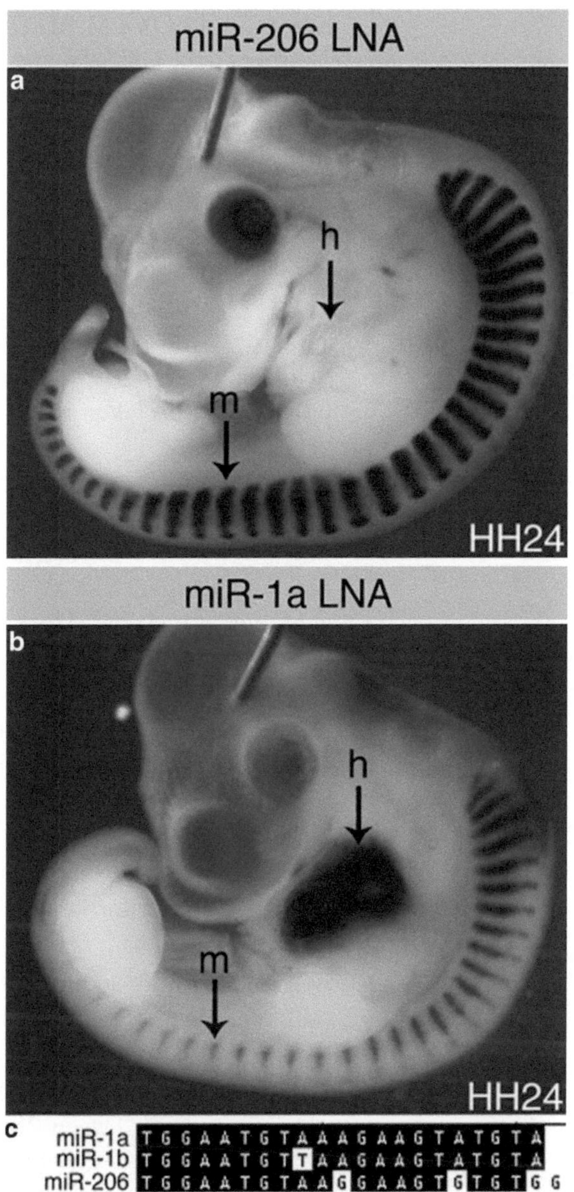

Fig. 2. Specificity of LNA probes. (**a**) LNA probe detects staining in skeletal muscle precursors muscle (m) of miR-206. (**b**) LNA probe detects both skeletal muscle precursors muscle (m) and heart (h) of miR-1a. (**c**) Sequence alignment shows high level of similarity between these probes.

3. PBSTw – DEPC-PBS with 0.1% Tween-20.
4. Hybridization buffer: 50% formamide, 1.3× SSC, pH 5, 5 mM EDTA, 50 μg/ml yeast RNA, 0.2% Tween-20, 0.5% CHAPS, and 100 μg/ml heparin. For 50 ml, add 3.25 ml of 20× SSC (pH to 5 with citric acid) to 25 ml of formamide.

Add 0.5 ml of 0.5 M EDTA, pH 8, 1 ml of 10% Tween-20, 2.5 ml of 10% CHAPS, and 17.5 ml of DEPC-treated water. Then add 125 μl of yeast RNA at 20 mg/ml (see Note 2) and 100 μl of heparin at 50 mg/ml. Store at –20°C.

5. Washing buffer: 50% formamide, 1× SSC, pH 5, and 0.1% Tween-20. Mix 25 ml formamide, 2.5 ml of 20× SSC (pH to 5 with citric acid), and 0.5 ml of 10% Tween-20. Make to a total volume of 50 ml and store at –20°C.
6. MABT: 100 mM maleic acid, 150 mM NaCl, and 0.1% Tween-20, pH to 7.5 with NaOH. A 5× stock can be made (without Tween-20), autoclaved, and diluted when required. Care should be taken when adjusting the pH as it is easy to overshoot.
7. Blocking solution: 1× MABT with 20% goat serum and 2% Roche Blocking Reagent (Roche). Heat to 65°C to dissolve Blocking Reagent. Alternatively, a 10% stock of Roche Blocking Reagent can be made in 1× MAB (without detergent) and autoclaved. This can be stored at –20°C and thawed when required.
8. Color reaction buffer (NMTT): 100 mM NaCl, 100 mM Tris–glycine, pH 9.5, 50 mM $MgCl_2$, and 1% Tween-20. Mix 1 ml of 5 M NaCl, 2.5 ml of 2 M Tris–glycine, pH 9.5, and 1.25 ml of 2 M $MgCl_2$; add 40.25 ml of H_2O and 5 ml of 10% Tween-20. Make fresh from stock solutions on the day of use.
9. NBT: 4-Nitro blue tetrazolium chloride, stock at 75 mg/ml in 70% dimethylformamide. Aliquot and store at –20°C. Protect from light.
10. BCIP: 5-bromo-4-chloro-3-indolyl-phosphate, 4-toluidine salt stock at 50mg/ml in dimethylformamide. Aliquot and store at –20°C. Protect from light.
11. 5× TBST: 0.68 M NaCl, 0.013 M KCl, 0.125 M Tris–glycine, pH 7.5, and 5% Tween-20. Add 40 g NaCl, 1 g KCl, and 125 ml 1 M Tris–glycine, pH 7.5, to 950 ml of dH_2O. Autoclave and then add 50 ml Tween-20.

3. Methods

1. Collect embryos and fix in 4% PFA at 4°C overnight. Any extraembryonic membranes should be removed with forceps in DEPC-PBS (see Note 1) and large cavities (such as brain vesicles) pierced.
2. Dehydrate embryos: wash in 50% methanol/PBSTw (see Note 3). Wash twice with 100% methanol. Leave at least

overnight at –20°C. Embryos can be stored for several weeks at this point.

3. Rehydrate embryos: wash in 75, 50, and 25% methanol/PBSTw. Wash twice in PBSTw. Ensure embryos have settled after each wash before proceeding.
4. Proteinase K – for younger embryos (inc *Xenopus*), this step can be omitted (see Note 4). For chicken embryos between HH25 and HH28, use 15 μg/ml of prot K; for embryos over HH28, use 20 μg/ml of prot K – incubate for 30 min at room temperature. Proteinase K is diluted in PBSTw for use. Rinse twice in PBST and then fix in 4% PFA/0.1% glutaraldehyde for 20 min at room temperature. Rinse and wash for 5 min in PBST.
5. Pre-hybridization: wash embryos in 1:1 PBSTw:hyb solution. Wash with hyb solution. Ensure embryos have settled before proceeding. Replace with hyb solution and incubate at hyb temperature (see Note 5) for at least 2 h before adding probe.
6. Hybridization: replace pre-hybridization with probe (see Notes 6, 7, and 8) prewarmed to the hyb temperature. Incubate overnight at hyb temp.
7. Post-hybridization washes: remove probe for reuse and rinse embryos twice in hyb buffer at hyb temperature. Then wash for 10 min in hyb buffer at hyb temperature and wash 3× for 1 h in washing buffer at hyb temp. Then wash overnight in washing buffer at hyb temp.
8. Wash 1× for 10 min at hyb temperature in 1:1 of washing buffer: MABT.
9. Rinse three times in MABT.
10. Wash twice for 30 min in MABT.
11. Replace with MABT/2% Roche Blocking Reagent, wash for 1 h.
12. Replace with blocking solution, wash for at least 1 h.
13. Replace with blocking solution containing a 1:2,000 dilution of anti-Dig-AP Fab fragments (Roche), and incubate overnight with rocking at 4°C.
14. Wash 5× 1 h in MABT, wash overnight in MABT at 4°C.
15. Color reaction: wash embryos twice for 10 min (minimum) in NMTT. Replace with NMTT with 9 μl NBT + 7 μl BCIP/ml (see Note 9). PROTECT FROM LIGHT. Continue color reaction until the background appears (generalized faint purple staining over the whole embryo) (see Note 10). At this point, wash twice for 10 min in 5× TBST, replace with fresh 5× TBST, and wash at 4°C overnight. Then the following day, repeat the color reaction until background appears and then wash in 5× TBST as before.

4. Notes

1. DEPC treatment of solutions. DEPC is used to destroy RNAse activity. After treatment, the solution must be autoclaved to remove the DEPC. DEPC will react with amines and so is not suitable for treating Tris. Only solutions used prior to hybridization need to be DEPC-treated as the duplex formed by the probe and its target is resistant to RNAse. Reagents such as Tris and EDTA, and detergents which cannot be DEPC-treated should be made in H_2O that has previously been DEPC-treated and autoclaved.
2. Several different suppliers and types of yeast RNA are available for use as a block in hybridization buffer. We have used Sigma yeast RNA (cat no. R6750) successfully. Stock solutions should be made using DEPC-treated water.
3. Washes: For all washes, allow at least 5 min/wash (unless otherwise specified in the protocol). For larger embryos, allow longer. It is important that the embryos are equilibrated with the solutions before moving on, so ensure that there is adequate time for this and a large enough volume (several times the volume of the embryos). In general, longer washes will not do any harm to the embryos. All washes are done at room temperature unless specified.
4. Proteinase K: LNA probes generally penetrate tissues well, so smaller embryos do not need treatment. Larger embryos (e.g., chicken HH 25 and older, mouse E13.5 and older) will require treatment. The figures given are a guideline – batches of PK can vary, so it may be necessary to optimize this. We use proteinase K from VWR cat no 390973P. Stocks are made at 10 mg/ml, stored at −20, and then diluted in DEPC-PBSTw just prior to use.
5. Hybridization temperature: This will need to be determined empirically for each probe. A good start is the predicted melting temperature −22°C; it may be necessary to try other temperatures as well. I have used temperatures including 42°C (miR-1a), 50°C (miR-206), and 65°C (miR-133). The same temperature should also be used for the post-hybridization washes.
6. Probes: LNA probes for in situ can be purchased from Exiqon. Although it is possible to buy unlabeled probes and end-label them with Dig-dd-UTP, much better results are obtained with pre-labeled probes. It is advisable to ask for Dig labeling on both the 3′ and 5′ ends of the probe to maximize the signal strength.

7. We have used LNA probes at 20 nM. Other groups have reported that 5 nM also works well (8). For a standard synthesis (10 μl at 25 μM), this means adding the probe to 12.5 ml of hyb solution. Probes can be stored in hyb solution at –20°C with no problems.
8. Pre-absorb probes before use. To do this, take embryos and perform a mock in situ up to the post-hybridization washes. Then discard the embryos. It is also a good idea to reuse probes as they will become cleaner with use. New probes always require a few uses before they become really good.
9. NBT/BCIP – as well as making from stock solutions, it is also possible to buy NBT/BCIP premixed or ready to use from Roche.
10. Color reaction: even good probes used several times rarely give a clean result to begin with. Usually, at first, no specific staining is apparent and a generalized purple staining appears over the whole embryo. At this point, it is necessary to wash the embryos in 5× TBST. This will remove the nonspecific staining and make it possible to continue the color reaction. Usually, this will have to be done several times before any specific staining is apparent, but repeated cycles of color/5× TBST can produce strong and specific staining. Embryos can be safely left in 5× TBST for days or even weeks, and in fact this can help clean them up even more. After a few washes in 5× TBST, the speed at which the background appears will slow, and so it will be possible to leave the embryos in color for longer before they need to be washed again. It is possible to leave the color reaction overnight once the rate of background formation has slowed but in this case, reduce the concentration of NBT/BCIP by half.
11. Double in situs: it is possible to combine a Dig-labeled LNA probe to detect a miRNA with a fluorescein-labeled antisense probe to detect a specific mRNA. In this case, the normal antisense probe should be hybridized at 65°C and then washed twice for 30 min at 65°C in wash buffer. Then the embryos should be returned to hyb buffer and then hybridized overnight with the LNA probe. Following this, the miR should be detected first as described above. Then, when the color is developed, the embryos should be fixed in 4% PFA with 0.1% glutaraldehyde overnight. Following this, the first antibody can be inactivated by incubating in PBSTw at 65°C for 1 h. Then the embryos can be blocked and incubated in anti-FITC antibody for detection with an alternative color.

References

1. Hornstein, E., Mansfield, J. H., Yekta, S., Hu, J. K., Harfe, B. D., McManus, M. T., Baskerville, S., Bartel, D. P., and Tabin, C. J. (2005) The microRNA miR-196 acts upstream of Hoxb8 and Shh in limb development *Nature* **438**, 671–4.
2. Shkumatava, A., Stark, A., Sive, H., and Bartel, D. P. (2009) Coherent but overlapping expression of microRNAs and their targets during vertebrate development *Genes Dev* **23**, 466–81.
3. Baek, D., Villen, J., Shin, C., Camargo, F. D., Gygi, S. P., and Bartel, D. P. (2008) The impact of microRNAs on protein output *Nature* **455**, 64–71.
4. Ason, B., Darnell, D. K., Wittbrodt, B., Berezikov, E., Kloosterman, W. P., Wittbrodt, J., Antin, P. B., and Plasterk, R. H. (2006) Differences in vertebrate microRNA expression *Proc Natl Acad Sci USA* **103**, 14385–9.
5. Braasch, D. A., and Corey, D. R. (2001) Locked nucleic acid (LNA): fine-tuning the recognition of DNA and RNA *Chem Biol* **8**, 1–7.
6. Vester, B., and Wengel, J. (2004) LNA (locked nucleic acid): high-affinity targeting of complementary RNA and DNA *Biochemistry* **43**, 13233–41.
7. Wienholds, E., Kloosterman, W. P., Miska, E., Alvarez-Saavedra, E., Berezikov, E., de Bruijn, E., Horvitz, H. R., Kauppinen, S., and Plasterk, R. H. (2005) MicroRNA expression in zebrafish embryonic development *Science* **309**, 310–1.
8. Darnell, D. K., Kaur, S., Stanislaw, S., Konieczka, J. K., Yatskievych, T. A., and Antin, P. B. (2006) MicroRNA expression during chick embryo development *Dev Dyn* **235**, 3156–65.
9. Sweetman, D., Rathjen, T., Jefferson, M., Wheeler, G., Smith, T. G., Wheeler, G. N., Munsterberg, A., and Dalmay, T. (2006) FGF-4 signaling is involved in mir-206 expression in developing somites of chicken embryos *Dev Dyn* **235**, 2185–91.
10. Sweetman, D., Goljanek, K., Rathjen, T., Oustanina, S., Braun, T., Dalmay, T., and Munsterberg, A. (2008) Specific requirements of MRFs for the expression of muscle specific microRNAs, miR-1, miR-206 and miR-133 *Dev Biol* **321**, 491–9.

Chapter 2

Detection of microRNAs in Plants by In Situ Hybridisation

Éva Várallyay and Zoltán Havelda

Abstract

MicroRNAs (miRNAs) are short, about 21 nucleotides in length, non-coding, regulatory RNA molecules representing a new layer in post-transcriptional gene expression regulation. Spatial and temporal analysis of miRNA accumulation by in situ analyses is the prerequisite of understanding the precise biological functions of miRNAs. Since miRNAs are very short molecules, their in situ analysis is technically demanding. Our method is based on the usage of highly sensitive LNA-modified oligonucleotide probes. LNA modification significantly enhances the sensitivity and specificity of miRNA detecting probes and provides relatively easy in situ miRNA detection. Here, we describe a protocol for this challenging technique step by step, in order to help every user to achieve success.

Key words: miRNA, Plant, LNA, In situ hybridisation

1. Introduction

miRNAs are key elements in biological processes; however, because of their small size, their presence and importance were discovered only with the rapid evolution of molecular biology techniques. To resolve their precise role, in situ spatial and temporal investigation of mature miRNA accumulation is needed. The technical problem that routs in the sort length of the target RNAs (21–25 nt) now seems to be solved with the use of locked nucleic acid (LNA)-modified probes (1). LNA modification in DNA oligonucleotides brings about a dramatically higher target affinity and specificity compared to a traditional DNA oligonucleotide (2). These probes have been introduced to enhance both the sensitivity and the specificity of miRNA detection by Northern blotting and in situ hybridisation (3–5). Using this technology, miRNAs can be detected relatively easily in both plant and animal tissues (5, 6). Here, we describe a detailed protocol for in situ

Tamas Dalmay (ed.), *MicroRNAs in Development: Methods and Protocols*, Methods in Molecular Biology, vol. 732,
DOI 10.1007/978-1-61779-083-6_2,

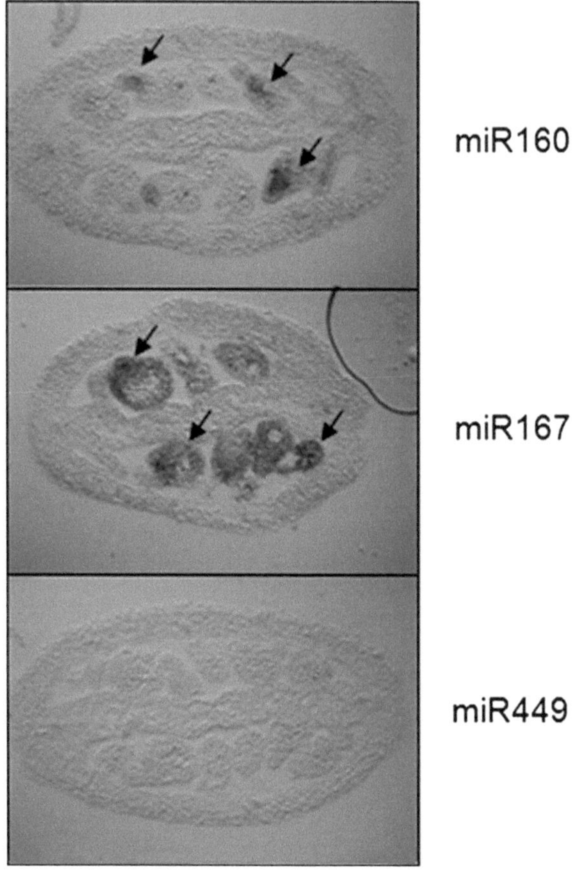

Fig. 1. Differential accumulation of miR160 and miR167 in *Arabidopsis thaliana* ovary. Longitudinal near-consecutive sections of *A. thaliana* ovary have been hybridised with 5′ and 3′ double-labelled LNA-modified oligonucleotides detecting miR160, miR167 and miR449 (chicken-specific miRNA is used as negative control). The hybridisation was carried out at 55°C overnight. *Arrows* show the accumulation of the miRNAs.

hybridisation of plant tissue sections using LNA-modified oligonucleotides (see Note 1) as probes (Fig. 1).

2. Materials

1. Alcian Blue solution: 0.25% in 3% acetic acid solution. Filtrate after preparation through a filter paper. Store at room temperature. You can use the same solution several times.
2. BCIP solution: 50 mg/ml in dimethylformamide. Store at –20°C.
3. Blocking solution I: Make it freshly every time. Dissolve 0.5% Blocking Reagent (Roche) in 1× TBS buffer at 60°C and cool down on ice.

4. Blocking solution II: Make it freshly every time. Dissolve 1% BSA in 1× TBS and add 0.3% Triton X-100.
5. DEI formamide: For hybridisation buffer, deionise with mixed bed resin (e.g., AG-501-×8 (Bio-Rad)) and store in aliquots at –20°C.
6. 100× Denhart's solution: Dissolve 10 g Ficoll400, 10 g polyvinylpyrrolidone and 10 g BSA in 500 ml sterile water. Keep aliquots frozen at –20°C.
7. Dextrane sulphate: 50% solution with water. Store in aliquots at –20°C. Note that it will dissolve only while boiling.
8. Fixative solution: Paraformaldehyde (Sigma; 4% w/v) and 0.1% Triton X-100 in phosphate-buffered saline (PBS) (see Note 2).
9. Hybridisation solution: 0.3 M NaCl, 10 mM Tris–glycine pH 6.8, 10 mM $NaHPO_4$ pH 6.8, 5 mM EDTA, 50% formamide, 10% dextran sulphate, 1× Denhardt's solution and 1 mg/ml tRNA.
10. NBT solution: 50 mg/ml in dimethylformamide. Store at –20°C.
11. 10× NTE buffer: 5 M NaCl, 100 mM Tris–glycine, pH 7.5, and 10 mM EDTA. Store at room temperature and dilute tenfold to have 1× working solution.
12. 10× PBS: 1.3 M NaCl, 0.07 M Na_2HPO_4 and 0.03 M NaH_2PO_4 (pH 6.5–7).
13. 20× Pronase buffer: 1 M Tris–glycine, pH 7.5, and 0.1 M EDTA. Store at room temperature and dilute it 20-fold to have 1× working solution.
14. Pronase solution: 40 mg/ml in water and self-digested at 37°C for 2 h to remove contaminant nuclease activities. Store in aliquots at –20°C.
15. RNase A solution: 10 mg/ml RNase A in water. Store in aliquots at –20°C.
16. 10× Saline: 8.5% w/v NaCl in water. Store at room temperature and dilute it tenfold to have 1× working solution.
17. 10× Salts buffer: 3 M NaCl, 0.1 M Tris–glycine, pH 6.8, 0.1 M $NaHPO_4$, pH 6.8, and 50 mM EDTA. Store at room temperature and dilute it tenfold to have 1× working solution.
18. 20× SSC: 3 M NaCl, 0.3 M Na_3 citrate. Store at room temperature and dilute it as indicated.
19. 10× Substrate buffer: 1 M Tris–glycine, pH 9.8, and 1 M NaCl. Store at room temperature.
20. 10× TBS buffer: 1 M Tris–glycine 1.5 M NaCl, pH 7.2.
21. tRNS solution: Prepare 100 mg/ml solution in water. Store in aliquots at –20°C.

3. Methods

In situ hybridisation of miRNAs depends upon the detection of RNA; therefore, it is very important to avoid contamination with RNases. The working environment should be clean and nuclease-free tubes, bottles, etc., should be used. The water and all aqueous solutions should be autoclaved and preferably aliquoted for single usage. (Because of its hazardous nature, we do not favour the use of diethyl pyrocarbonate treatment.)

3.1. Fixation and Embedding (Table 1)

Embedding of plant tissues in vax is a long and slow process that takes several days. After fixation of the plant tissues, we slowly saturate the tissue parts with wax to make them solid for microtom cutting. For fixation, we use formaldehyde solution, but it is not possible to change this solution directly with wax. So, first we change formaldehyde to ethanol in order to dehydrate tissues and then replace ethanol with histoclear which is a good solvent for wax. Next, we saturate the tissue parts with wax. The time of fixation is the most important step of the protocol, because if the fixation is not complete we can lose the signal and the tissue structure. On the contrary, if we over-fix the material, it will be difficult for the probe to penetrate the tissue, resulting again in the loss of signals. So, we have to optimise the fixation time for our tissue of interest. We usually refer to parameters what we have used for *Arabidopsis thaliana* or *Nicotiana benthamiana* tissues.

1. Remove samples of your interest from the plant, if it is bigger than 1 cm, then cut them into half and place them immediately into ice-cold fixative solution. We usually use 15 ml plastic tubes without the cap as container (see Note 3).
2. Keep the tubes on ice, weigh the tissue under the surface of the fixative and apply vacuum in a chamber till bubbles are formed (usually 20–30 s with the pump working) and hold the vacuum for 5–10 min. Release the vacuum very slowly. Since formaldehyde is toxic, work under a fume hood.
3. Repeat the vacuum treatment and change the fixative solution.
4. Repeat the whole procedure (2–3) until all the samples sink to the bottom of the tube (this is the indication of complete infiltration of fixative solution into the tissue samples).
5. Exchange the fixation solution when all the samples have sunk down and fix the samples at 4°C for 4 h to overnight with gentle shaking (see Note 4).
6. Degas the 1× saline solution under vacuum before usage. Wash the fixative solution with ice-cold 1× saline solution on ice.

Table 1
Fixation and embedding

Fixative solution	on ice	20–30 s	Under vacuum, several times
Fixative solution	on ice	O/N	Gentle shaking
1× saline	on ice	30 min	Degass under vacuum before usage
30% EtOH/saline	on ice	1–3 h	Degass under vacuum before usage
40% EtOH/saline	on ice	1–3 h	Degass under vacuum before usage
50% EtOH/saline	on ice	1–3 h	Degass under vacuum before usage
60% EtOH/saline	on ice	1–3 h	Degass under vacuum before usage
70% EtOH/saline	on ice	1–3 h	Degass under vacuum before usage
85% EtOH/saline	4°C	1–3 h	Degass under vacuum before usage
95% EtOH/water	4°C	1–4 h	Degass under vacuum before usage
100% EtOH	4°C	1–4 h	
100% EtOH	4°C	O/N	
100% EtOH	RT	1–2 h	
100% EtOH, 0.1% Eosin	RT	30 min	
EtOH/Histoclear = 3:1	RT	1–3 h	
EtOH/Histoclear = 1:1	RT	1–3 h	
EtOH/Histoclear = 1:3	RT	1–3 h	
Histoclear	RT	1 h	
Histoclear	RT	O/N	
Histoclear + WAX	RT	~1 day	Add wax chips several times
Histoclear + WAX	42°C	2 days	Add wax chips several times
WAX	58°C	2 days	Change half volume of the wax 2–3 times a day
WAX	58°C	2 days	Change total volume of the wax 2–3 times a day
WAX	4°C		After orientation store for unlimited time

7. Dehydrate the samples by passing them through a graded ethanol series (30, 40, 50, 60, 70 and 85% EtOH/1× saline, followed by 95% EtOH/water and finish in 100% EtOH). Degas the EtOH/1× saline-graded ethanol series under vacuum before usage. For detailed description of solution and temperature, refer to Table 1. You can interrupt the protocol and leave the samples overnight at any step after 50% EtOH/1× saline (see Note 5).

8. The next steps are performed at room temperature with gentle shaking. Replace 100%EtOH with fresh one and leave it for 1–2 h. During this period, the green colour must be eliminated. If not, repeat this step until the tissues become colourless. Then, change the EtOH with 100% EtOH containing 0.1% EosinY and agitate the tubes for 30 min (see Note 6).
9. Pass the samples through a graded histoclear series with gentle rotation (refer to Table 1. for detailed description). It is advised to carry out the Histoclear steps in glass vial or good quality plastic reaction tube, otherwise Histoclear may damage the tube. Depending on the size of the samples, every step takes 1–3 h.
10. Add one Paraplast chip to about 1 ml of histoclear and leave it to dissolve at room temperature. Once the first chips have dissolved add other ones, and so on, until the chips will not dissolve any more. At this stage, transfer the samples to 42°C and saturate the histoclear by adding more and more Paraplast chip till saturation.
11. Put samples at 58°C and overlay it with freshly melted Paraplast (avoid heating Paraplast over 58°C). Try to form a cap of melted wax on the top of solution to avoid heat stress of the samples. Change half of the volume for freshly melted Paraplast and leave at 58°C overnight.
12. Change half of the wax twice a day to freshly melted wax for 2 days. (Decant gently the old Paraplast and pour in fresh melted Paraplast.) Leave the tubes open to allow Histoclear to evaporate.
13. Try to change the total volume to freshly melted wax for another 2 days.
14. Prior to sectioning, the tissue samples must be solidified in a wax block. Put the plastic or metal moulds on a hot plate at 60°C, shake up the samples and pour them out into the mould. Ideally, about 10–12 samples should be transferred into one mould. Arrange the samples in rows leaving sufficient space between them allowing the cutting out of single blocks containing one or a few samples. Put the samples on ice or at 4°C and allow them to harden. Apply the tissue block on a holder compatible with the microtome.

3.2. Sectioning

Good sections are very important for nice results. It is worthwhile to practise it with somebody who has done it before. You will need patience and practise and correctly embedded samples to do perfect sections.

1. Make tissue sections (8–20 μm) using a retracting rotary microtome. Trim the block so that the upper and lower faces are parallel. Repeated sectioning leads to the formation of

ribbon of sections. Trimming a trapezoid shape block helps the identification of a single section in the ribbon.

2. Mount the sections onto poly-L-lysine-coated pre-prepared slides. Wax sections need to be stretched before adhesion to the glass slide. Sections should be put onto a layer of degassed water on a slide held on a warmed hotplate (40–42°C). Once the section has stretched, drain away the excess water and leave the slide to dry overnight 40–42°C.

3.3. LNA Probe Preparation and Probe Checking

LNA-modified oligonucleotide-based probes can be purchased from Exiqon (Denmark; www.exiqon.com) and its labelling depends on the required sensitivity. *Arabidopsis thaliana* tissue sections respond less efficiently in in situ experiments than other plant tissues (for example *N. benthamiana*). While the 3′ end-labelled probes (using the DIG oligonucleotide 3′-end labelling kit (Roche)) work well in *N. benthamiana*, they do not provide reliable signals in *A. thaliana*. To achieve good signals in *A. thaliana* in situ experiments, it is recommended to order 5′ chemically DIG-labelled LNA-modified probe and also label at the 3′ end, producing a double-labelled probe. Alternatively, double-labelled probes are also possible to order. It is very important to use a similarly labelled negative control (for example, an animal-specific miRNA) during the experiments to monitor background hybridisation.

1. Label 50 pmol LNA-modified oligonucleotide or 5′ DIG-labelled LNA-modified oligonucleotide in 10 μl volume using the DIG oligonucleotide 3′-end labelling kit (Roche), according to the manufacturer's instructions. It is not necessary to purify the probe after labelling.
2. Remove 0.2 μl (for probe checking) from the 10 μl reaction and add 10 μl deionised formamide.
3. It is sensible to check the quality of the labelled probe prior to usage. To do this, spot 0.2 μl aliquot (and the labelled control oligonucleotide provided by the kit) on a piece of membrane and cross-link with UV light.
4. Put the membrane in 1× TBS buffer for 2 min and incubate the membrane in TBS buffer containing 5% of powdered milk for 10 min (Blocking). Add anti-DIG-alkaline phosphatase and Fab fragments (1:2,000) and hybridise with gentle shaking for 30 min.
5. Wash at least three times in 1× TBS buffer for 5 min each and transfer to 1× SB for 2 min (washing).
6. Develop the colour reaction by adding NBT and BCIP (add 3 μl NBT and 3 μl BCIP solution into 1 ml 1× SB). Stop the reaction by rinsing the membrane with water and dry the membrane.

3.4. Slide Preparation for Hybridisation (Table 2)

To make the tissue sections on the slide accessible for solvents, you have to remove wax. During the protocol, you have to move the slides together between different solvents. You can decide to move them individually (it is very difficult to handle more than 10 slides with this method) or you can put them in a slide holder. (This is more convenient and faster. With this method, you can move up to 20 slides together.)

1. Transfer the slides into histoclear at room temperature and incubate for 10 min. For the first treatment, the histoclear can be reused from the second treatment of a previous experiment. Repeat using fresh histoclear.
2. Transfer slides into 100% EtOH (can be reused from the previous experiment before) for 5 min. Repeat once using fresh 100% EtOH.
3. Hydrate the samples by passing them through a graded ethanol series (95, 85, 75, 50, 30% EtOH/1× saline) for 2 min at each step. For a detailed description of solutions and temperature, refer to Table 2. This ethanol series can be reused for several times in the subsequent experiments, and can be reused directly in the reverse order in this experiment at a later stage (see point 6). Finish by washing the sides in 1× saline for 2 min.
4. Equilibrate the slides in 1× pronase buffer for 2 min at 37°C. Incubate the slides in pronase solution (containing 10 μg/ml pronase) for 10 min at 37°C. Incubate the slides in 0.2% glycine in 1× PBS for 2 min and then wash them in 1× PBS for 2 min.
5. Postfix in fixative solution (in a fume hood) for 30 min. Wash the slides twice in 1× PBS for 2 min each.
6. To eliminate background reaction of the hybridisation probe, due to electrostatic binding, amino groups on the section should be acetylated using an acetic anhydride treatment (1 ml acetic anhydride in 200 ml 0.1 *M* triethanolamine-HCL, pH 8.0). Incubate the slides in buffered acetic anhydride for 10 min at room temperature (see Note 7).
7. Wash the slides once in 1× PBS for 2 min. Dip the slides in fresh 1× saline solution for 2 min and dehydrate through a graded ethanol/saline series (30, 50, 75, 85, 95% EtOH/1× saline) for 2 min at each step. If you have saved the ethanol series at point 3, you can reuse it now.
8. Transfer the slides to 100% EtOH.

Now the slides are ready for hybridisation. You can stop here and keep the slides safely in EtOH for hours, but in this case put them to 4°C.

Table 2
Slide preparation for hybridisation

(a) Removal of the wax from the tissues			
Histoclear	10 min	You can use a used one, but waste it after usage	
Histoclear	10 min	Use fresh one and keep it for reuse once	
100% EtOH	5 min	You can use a used one, but waste it after usage	
100% EtOH	5 min	Use fresh one and keep it for reuse	
95% EtOH/saline	2 min		
85% EtOH/saline	2 min		
75% EtOH/saline	2 min		
50% EtOH/saline	2 min		
30% EtOH/saline	2 min		
100% 1× saline	2 min		
All these reactions should be made at room temperature			
(b) Treatment with pronase and elimination of background signals			
1× Pronase buffer	37°C	2 min	
Pronase in pronase buffer	37°C	10 min	50 μl 40 mg/ml Pronase solution in 200 ml pronase buffer
Glicin	RT	2 min	0.2% Glycine solution in 1× PBS
PBS	RT	2 min	
Fixative solution	RT	30 min	
PBS	RT	2 min	
PBS	RT	2 min	
Acetic anhydride	RT	10 min	For 200 ml solution add 1 ml acetic anhydride to 2.5 ml triethanolamine + 1 ml ccHCl to sterile water just before using
PBS	RT	2 min	
100% 1× saline	RT	2 min	
30% EtOH/saline	RT	2 min	
50% EtOH/saline	RT	2 min	
75% EtOH/saline	RT	2 min	
85% EtOH/saline	RT	2 min	
95% EtOH/saline	RT	2 min	
100% EtOH	RT	2 min	If you leave the slides for longer period in EtOH, put to 4°C

3.5. Hybridisation

1. Prepare hybridisation solution about 100–200 μl per slide depending on the number and size of the sections. For 1 ml hybridisation solution, add 100 μl 10× salts buffer, 500 μl deionised formamide, 250 μl 50% dextran sulphate, 10 μl 100 mg/ml tRNA, 10 μl 100× Denhardt's solution and water (see Note 8).
2. Add 1–5 μl labelled LNA probe (in 50% formamide) per slide to the hybridisation solution. Vortex well, centrifuge to eliminate bubbles and keep at the temperature of hybridisation (50–60°C) (see Note 9).

3. Put one slide on a hot plate at 50°C and allow it to dry. Do not over dry your sections. Apply the hybridisation solution with probe as a band along the middle of the slide and cover it. Put the slide in a closed environment saturated with 50% formamide/2× SSC pre-warmed to the temperature of hybridisation. Prepare a plastic box with 50% formamide/ 2× SSC at bottom. Place the slides on a horizontal support inside the plastic box (see Note 10).
4. Repeat with every slide individually. Close the box, seal with clingfilm and incubate the slides at the temperature of hybridisation overnight. Prepare washing solution in excess (0.2× SSC) and place at the temperature of hybridisation.

3.6. Washing and Signal Detection (Table 3a, b)

1. Start washing at 50–60°C (the same temperature as hybridisation) in 0.2× SSC. Put slides into washing solution and carefully remove the cover slips or parafilms. Rinse the slides having different probes separately several times to avoid cross-contamination of probes and wash them for 30 min. Wash the slides twice at the temperature of hybridisation for 1 h each (see Note 11).
2. Immerse the slides in 1× NTE buffer pre-warmed to 37°C, repeat in fresh buffer 5 min each. Incubate the slides in NTE containing 20 μg/ml RNase A at 37°C for 30 min to remove background hybridisation.
3. Rinse the slides in 1× NTE for 5 min at room temperature and transfer them to washing solution (0.2× SSC) for 1 h at the temperature of hybridisation. Dip the slides into 1× SSC for 2 min then into 1× TBS twice for 5 min each time. Slides are now ready for the detection step.
4. Incubate the slides in Blocking solution I for 30 min. Transfer the slides into Blocking solution II and incubate for 30 min (see Note 12).
5. Add anti-DIG-alkaline phosphatase, Fab fragments (Roche) (1:2,000) to the required amount of Blocking solution II. (0.5 ml per slide). Place the slides individually on a horizontal support and put them into moist plastic or glass chamber on a tray. Apply the antibody solution onto the slides and incubate for 90 min at room temperature.
6. Stop the reaction by washing the slides (transferred back to the rack) at least five times in excess 1× TBS for 5 min. Equilibrate the slides in 1× SB for 5 min.
7. To develop the colour reaction, apply NBT and BCIP containing 1× SB (add 30 μl NBT and 30 μl BCIP solution to 10 ml 1× SB) on the slide. Remove the slides one by one from 1× SB and immediately apply the substrate solution because after drying it can be difficult to spread the liquid. Put the

Table 3
Washing and signal detection

(a) Washing			
0.2× SSC	Temperature of hybridisation	30 min	Wash slides with different LNA probes separately
0.2× SSC	Temperature of hybridisation	1 h	
0.2× SSC	Temperature of hybridisation	1 h	
NTE	37°C	5 min	
NTE	37°C	5 min	
RNase A in NTE	37°C	30 min	400 μl 10 mg/ml RNase A solution in 200 ml NTE
NTE	RT	5 min	
0.2× SSC	Temperature of hybridisation	1 h	
1× SSC	RT	2 min	
TBS	RT	5 min	
TBS	RT	5 min	
(b) Blocking, hibridisation with anti-DIG and staining			
Blocking reagent/TBS	30 min	1 g blocking reagent/200 ml 1× TBS	
1% BSA, 0.3% Triton X-100/TBS	30 min	2 g BSA/600 μl Triton X-100/in 200 ml TBS	
1% BSA, 0.3% Triton X-100/TBS + Anti-DIG 1:2,000	90 min	Slides individually 500 μl/slide	
TBS	5 min		
TBS	5 min		
TBS	5 min		
TBS	5 min		
TBS	5 min		
Substrate buffer	5 min	Dilute 20 ml 10× SB + 4 ml 2.5 M $MgCl_2$ to 200 ml with distilled water	
NBT/BCIP/$MgCl_2$/substrate buffer	2–24 h and up	30 μl NBT and 33 μl BCIP in10 ml in $MgCl_2$/SB Slides individually 0.8–1 ml/slide	
All these reactions should be made at room temperature			

(continued)

Table 3 (continued)

(c) Final washing and background staining		
Distilled water	2–5 min	
40% EtOH/water	2–5 min	
70% EtOH/water	2–5 min	
100% EtOH/water	2–5 min	
70% EtOH/water	2–5 min	
40% EtOH/water	2–5 min	
Distilled water	2–5 min	
Alcian Blue	5–15 min	99 ml water + 1 ml acetic acid +0.25 g Alcian blue
Distilled water	2 min	
All these reactions should be made at room temperature with gentle shaking		

slides into a moisture chamber and cover them individually with about 0.8–1 ml of substrate solution. Cover the moisture chamber to keep the slides dark, until staining develops.

8. Monitor the signal development for 2–24 h. For rare miRNA targets, you may need longer signal development time. Stop the reaction at the desired signal intensity by rinsing the slides in water (see Note 13).

3.7. Final Washing and Background Staining (Table 3c)

Before doing this final washing step, investigate your samples in order to decide if it is necessary or not. Compare the slides with negative control and if you detect high level of background hybridisation, then you need to wash your slides. If you don't need final washing, go directly to the counterstaining step (point 2).

1. Wash the slides in a graded EtOH series for 2–5 min (depending on the intensity of the signal and background) at each step (40, 70 and 100% EtOH). Repeat the process in reverse direction. For a detailed description of solution and temperature, refer to Table 3.
2. Counter-stain the sections by dipping the slides for 5–15 min in 0.25% Alcian blue in 3% acetic acid. The slides should be monitored for the intensity of staining (tissue having no hybridisation signals should show a faint blue staining). Rinse the slides in water and air-dry.
3. Cover the slides with cover slips using mounting solution (DPX), about 100–200 μl per slide. Leave the slides to dry for a few hours. Now the sections are ready for data recording using a standard light microscope (see Note 14).

4. Notes

1. LNA-modified oligonucleotide probes detecting miRNAs can be ordered from Exiqon (http://www.exiqon.com) and a website for probe design is also available (http://lnatools.com).
2. It should be prepared in a screw-top bottle (e.g., Duran type) under a fumehood. Take 100 ml 1× PBS, pH = 6.5–7 and using a solution of NaOH adjust pH = 11. Measure 4 g paraformaldehyde to the bottle, fill it with the 1× PBS and heat the solution to 60°C. Shake vigorously for about 30 s and release the pressure every 5–10 s. Leave to settle. Most of the paraformaldehyde should be dissolved, though the solution may still be slightly cloudy. Cool it on ice. Use H_2SO_4 to adjust the pH back down to 7 (do not use HCl as this releases a carcinogen). Add 0.1 ml of Triton X-100 to the solution. You can

prepare larger volume solution and store it at –20°C in aliquots. Once you have thawed an aliquot do not freeze back.

3. The ratio between the sample and the fixative solution is crucial. In a 15-ml tube, in 8–10 ml fixative put no more than 10 pieces of 1 × 1 cm tissue from bigger leaves, 20–30 *Arabidopsis* seedling, 20–30 *Arabidopsis* siliques and 20 *Arabidopsis* inflorescens. For samples such as silique, make a cut at some part to make it possible to the fixative to get inside.
4. The fixation time strongly depends on the size and the tissue type of the samples. Larger and compact tissue samples usually require longer fixation (for example, 4h for *Arabidopsis* seedling, 6–8 h *Arabidopsis* inflorescence and 12–14 h for *Arabidopsis* siliques); however, over fixation can reduce the hybridisation signal.
5. Depending on the size of the samples, every step takes 1–3 h. For *Arabidopsis* leaves, inflorescence the above protocol is perfect. However, for more compact tissues such as *Nicotiana* leaves and seeds, it is more sensible to use less steps, but for longer incubation time (50–70–85–95%EtOH/1× saline for up to 3 h every step).
6. Staining with eosin is important as it will make orientation and sectioning easy. Without eosin staining, you won't see the tissue in the wax. Do not over stain your samples, as it will be difficult to wash out this stain from the tissue sections and it may interfere with the signal detection.
7. To prepare 200 ml 0.1 *M* triethanolamine-HCL buffer, pH 8.0, add 2.5 ml triethanolamine and 1 ml HCl to water and stir. Add 1 ml acetic anhydride to 200 ml triethanolamine buffer and stir vigorously (work under fume hood). Since the acetic anhydride is very unstable in water, it has to be added just before using.
8. The volume of the probe usually does not alter significantly the concentration of hybridisation solution. As this hybridisation buffer is very viscous, prepare a bit more than the desired volume as you can lose quite a lot of solution with pipetting.
9. The temperature of hybridisation strongly depends on the nature of the particular probe. 55°C is a good starting temperature. If the probe tends to give background hybridisation, then increase the temperature of hybridisation to 60°C and in parallel increase also the temperature of washing. If no signal is detected, then lower the temperature of hybridisation and washing to 50°C.
10. The slide must be covered here, but with a traditional cover slip it seemed very difficult for us. As an alternative, cut Parafilm precisely into a coverslip size and with the help of forceps, put the Parafilm on the slide from one side of the

slide slowly. With this method, formation of bubbles under the cover slip can be prevented. It is very important to prevent the bubble formation because hybridisation will not occur in the presence of bubbles.

11. Be very cautious with the removal of the cover slips as you can harm the tissue sections easily. The easiest way to this is to put the slide vertically into washing buffer and wait a bit until the cover slips fall off.
12. Blocking reagent is difficult to dissolve in 1× TBS at room temperature. You have to heat it, but before usage, put the solution on ice because ice-cold Blocking solution gives much better result than room temperature solution.
13. If you have to wait for signal development for a longer period, change the NBT/BCIP substrate solution twice a day. We usually do not cover the slides during stain development, so it is easy to do this; just drain off the used stain and apply new one on the slide. Although it is possible to check the staining with the stains on, in this case, you have to cover the slides with a cover slip.
14. After using DPX, it is very difficult to remove the cover slips (you can remove them by putting the slides in histoclear). Use 50% glycerol if you want to further wash your slides.

Acknowledgments

This work was supported by a grant from the Hungarian Scientific Research Fund (OTKA K78351). VÉ is a recipient of Bolyai Janos Fellowship.

References

1. Kauppinen, S., Vester, B., and Wengel, J. (2006) Locked nucleic acid: high-affinity targeting of complementary RNA for RNomics. *Handb Exp Pharmacol* **173**, 405–22.
2. Valoczi, A., Hornyik, C., Varga, N., Burgyan, J., Kauppinen, S., and Havelda, Z. (2004) Sensitive and specific detection of microRNAs by northern blot analysis using LNA-modified oligonucleotide probes. *Nucleic Acids Res* **32**, e175.
3. Varallyay, E., Burgyan, J., and Havelda, Z. (2007) Detection of microRNAs by Northern blot analyses using LNA probes. *Methods* **43**, 140–45.
4. Varallyay, E., Burgyan, J., and Havelda, Z. (2008) MicroRNA detection by northern blotting using locked nucleic acid probes. *Nature Protocols* **3**, 190–96.
5. Valoczi, A., Varallyay, E., Kauppinen, S., Burgyan, J., and Havelda, Z. (2006) Spatio-temporal accumulation of microRNAs is highly coordinated in developing plant tissues. *Plant Journal* **47**, 140–51.
6. Kloosterman, W. P., Wienholds, E., de Bruijn, E., Kauppinen, S., and Plasterk, R. H. (2006) In situ detection of miRNAs in animal embryos using LNA-modified oligonucleotide probes. *Nat Methods* **3**, 27–9.

Chapter 3

Detecting sRNAs by Northern Blotting

Sara López-Gomollón

Abstract

Small RNAs (sRNAs) play fundamental roles in modifying the transcriptomes and proteomes of organisms. sRNAs can be classified according to their origin and way of action into classes such as microRNAs, *trans*-acting siRNAs, heterochromatin siRNAs, Piwi-interacting RNAs, and natural antisense siRNAs, although microRNAs are the most conserved and studied class. The expression pattern of a sRNA can be indicative of its biological function. Northern blotting is one of the most widely used methods to identify, validate, and study the expression profile of sRNAs because it is a quantitative, relatively inexpensive technique that is readily available for most laboratories. This chapter describes a protocol for sRNA Northern blot analysis, which includes RNA extraction, polyacrylamide gel electrophoresis, hybridisation, or the detection of sRNA using oligonucleotide probes.

Key words: Short RNAs, MicroRNAs, Gene expression regulation, Gene silencing, RNAi, RNA silencing, Northern blot, Transference, Hybridisation, Nucleic acid detection

1. Introduction

Gene expression is regulated at several levels, which is crucial for coordinating cellular pathways. Recently, a post-transcriptional mechanism was discovered that involves small RNA (sRNA) molecules that produce the sequence-specific inhibition of gene expression (1). Most sRNAs are derived from double-stranded RNA (dsRNA) that is processed to 21–24 nt dsRNA with 3′-overhanging ends. One of the strands of dsRNA is degraded, allowing the other to pair with homologous mRNA in an Argonaute (Ago) protein-mediated process to direct post-transcriptional repression (2, 3). The sRNAs are classified depending on their origin and way of action. In plants microRNAs (miRNAs) are generated from a precursor RNA showing hairpin structures by Dicer-like 1 (4). Other sRNAs are produced from dsRNA

Tamas Dalmay (ed.), *MicroRNAs in Development: Methods and Protocols*, Methods in Molecular Biology, vol. 732, DOI 10.1007/978-1-61779-083-6_3,

synthesised by RNA-dependent RNA polymerase 6 (*trans*-acting siRNAs) (5, 6), by RDR2 (heterochromatin siRNAs) (7), or by overlapping antisense mRNAs (natural antisense siRNAs) (8).

Analyses of sRNAs by Northern blot provide not only valuable information about expression profiles or abundance but also about the size of the sRNAs. One of the most crucial steps is to obtain a high-quality RNA from the samples maintaining the sRNA fraction. After that, it is necessary to separate RNAs by size using polyacrylamide gel electrophoresis (PAGE), stabilise them onto a solid support by transferring the RNA onto a membrane and bind them covalently through a cross-linking reaction. After that, the expression of sRNAs can be assessed by hybridisation with an adequate DNA oligonucleotide probe. This chapter describes the optimised methods for all the steps required to study sRNA expression using Northern blotting.

2. Materials

2.1. RNA Extraction

1. TRIzol reagent (Invitrogen Life Technologies, Paisley, UK). It is a monophasic solution of phenol and guanidine isothiocyanate. It is toxic in contact with skin and if swallowed and can cause burns. It must be kept below room temperature and in a dark-coloured container covered in foil because it is light sensitive (9).
2. Chloroform.
3. Isopropyl alcohol.
4. 75% Ethanol in RNase-free water.
5. RNase-free water. To prepare it draw water into RNase-free glass bottles. Add diethylpyrocarbonate (DEPC) to 0.01% (v/v). Incubate in a fume hood at room temperature overnight. Autoclave to inactivate the residual DEPC and cool before use. It can be stored at room temperature. DEPC irritates the eyes, respiratory system, and skin.
6. 5 M NaCl.
7. 10% SDS. SDS causes irritation to the respiratory tract. Use a fume hood or a mask when weighing the powder.
8. 1.95% (w/v) Na_2SO_3.
9. 2-Mercaptoethanol.
10. 10× TBE. 890 mM Tris base, 890 mM boric acid, and 20 mM EDTA [108.0 g Tris base, 55 g boric acid, and 40 ml 0.5 M EDTA (pH 8.0)]. Bring to 1 L with DEPC-treated H_2O, autoclave, and store at room temperature.
11. Tris-saturated phenol (pH 8.0).
12. Chloroform–isoamyl alcohol (24:1).

2.2. Denaturing PAGE

1. 50 ml 15% denaturing polyacrylamide gel: Mix 25 g urea, 5 ml 10× TBE or 10× MOPS, 18.75 ml 40% acrylamide (38:2 = acrylamide:bis-acrylamide), and 10 ml sterile water. Dissolve the urea in water by heating in a microwave oven but not let it boil. When the urea is completely dissolve, add polyacrylamide and TBE or MOPS. Add sterile water up to 50 ml. Just before pouring the gel, add 300 μl of 10% ammonium persulphate (APS) and 20 μl of TEMED and pour the gel immediately. Add 5 ml of 1M sodium acetate and 5 ml of DEPC-treated 0.5 M EDTA (pH 8.0). Polymerisation time is about 30 min.
2. 10× MOPS/NaOH pH 7.0 (200 mM) buffer. To make 250 ml, dissolve 11.56 g of MOPS in 200 ml of DEPC-treated H_2O. Adjust the pH to 7.0 with NaOH. Make up to 250 ml with DEPC-treated H_2O. Filter sterilise (do not autoclave) and cover with foil. Light yellow (straw-like) coloured is still usable but discard if it turns intensively yellow.
3. 2× Loading buffer: 5 mM EDTA, 0.1% bromophenol blue, 0.1% xylene cyanol, and 95% formamide.
4. 10 mg/ml Ethidium bromide (EtBr). It is thought to act as a mutagen because it intercalates double-stranded DNA, thereby deforming the molecule.

2.3. RNA Transfer

1. 10 mM Na-phosphate buffer (pH 7.0): Dissolve 0.58 g of monosodium phosphate monohydrate and 1.55 g of disodium phosphate heptahydrate in 1 L of DEPC-treated H_2O.
2. 20× SSC. 3 M sodium chloride and 300 mM trisodium citrate. (175.32 g NaCl, 88.23 g trisodium citrate dihydrate. Adjust to pH 7.0 with HCl. Bring to 1 L with DEPC-treated H_2O, autoclave, and store at room temperature).

2.4. Cross-linking

1. Cross-linking solution: Add 122.5 μl 12.5 M 1-methylimidazole to 10 ml ddH_2O. Adjust pH to 8.0 by the addition of 1 M HCl (usually requires 150 μl). This can be prepared 1–2 h before use and kept at room temperature. Weigh 0.373 g l-ethyl-3-(3-dimethylaminopropyl) carbodiimide (EDC) and dissolve the EDC by the addition of the 1-methylimidazole solution. Add ddH_2O up to 12 ml. This provides a working solution of 0.16 M EDC in 0.13 M 1-methylimidazole at pH 8.0 and is sufficient to saturate two 6×9 cm Whatman papers.

2.5. Hybridisation

1. γ-^{32}P-ATP. ^{32}P is a source of radio hazard. Its use requires user training and strict workplace safety regulations.
2. Wash buffer: 0.2× SSC and 0.1% SDS.
3. Stripping solution: 10 mM Tris–glycine, pH 8.5, 5 mM EDTA, and 0.1% SDS. Use a fume hood or a mask when weighing SDS.

3. Methods

The first step to detect sRNA is to extract RNA from the samples. It is important to choose a method suitable for your samples to yield high-quality RNA. Column-based kits are not recommended to extract RNA because the sRNA fraction binds to the column weakly, and it can be lost during washing. Phenol-based extraction methods, such as the one described later, are popular because they produce high yields of total RNA, although these techniques do require the use of toxic chemicals. In any protocol used, it is also important to avoid LiCl precipitations because of the low recovery of sRNAs. An improved method is needed to extract RNA from tissues with high lipid, proteoglycan, and polysaccharide content. This chapter also provides a method to extract RNA from plant tissues with high levels of polysaccharides and polyphenols, such as tomato fruits (10). There are commercial kits such as miRVana (Ambion, UK) or miRNeasy kit (Qiagen, UK), where the resulting RNA preparation is highly enriched for sRNAs, allowing more sensitive sRNA detection compared with the same assay using total RNA. Nevertheless, the use of sRNA-enriched samples is not encouraged if the comparison of sRNA expression among different RNA samples is desired. This is because the gel is equally loaded compared with the sRNA samples and the percentage of the sRNA fraction in each RNA sample can be different.

Samples are separated on a denaturing polyacrylamide gel that offers enough resolution to differentiate sRNAs from other short RNAs, such as tRNAs or small nucleolar RNA. The percentage of the gel should be adapted to the running system. We usually use the Bio-Rad Mini Protean II system (Bio-Rad, UK) and run 15% gels. After running the gel, the RNA is transferred to a membrane. Two different methods are described in this chapter to transfer RNA to a membrane. The capillary method described by Southern in 1975 (11) is still a widely used technique. One of its main advantages is that no equipment is required. The second method uses a semidry system, which is quicker than the capillary method. After blotting the gel, it is necessary to cross-link the RNA to the membrane to ensure it is covalently bound to the membrane. It can be done using either a UV source or a chemical cross-linker. UV is faster but less efficient for sRNAs than for mRNAs because the nucleotides cross-linked to the membrane cannot participate in the annealing to the probe. An alternative protocol is also described here, where sRNAs are linked to the membrane only through their 5′-terminal phosphates (12). Please note that the cross-linking method determines the electrophoresis buffer and membrane.

Finally, the RNA blot is hybridised using a DNA oligonucleotide probe. The probe is complementary to the short RNA it is intended to detect, and is 5′-radiolabelled with γ-^{32}P-ATP. We can increase the sensitivity and specificity of the sRNA detection using a locked nucleic acid (LNA) probe. LNA nucleotide has a ribose modified with an extra bridge connecting the 2′ and 4′ carbons. The bridge "locks" the ribose in a 3′-endo structural conformation, which enhances base stacking and backbone pre-organisation. This increases significantly the thermal stability (13). After hybridisation, the membrane is washed and exposed to either X-ray film or a phosphorimager plate, and the signal is visualised. The membranes can be stripped and used again.

3.1. RNA Extraction

3.1.1. TRIzol Extraction

1. Homogenisation (see Note 1):

   Tissues: Place sterile mortals and Eppendorf tubes on ice. Grind your sample in the ice-chilled mortar into a fine powder. Add 1.5 ml of TRIzol to 50–100 mg of the sample and mix by vortexing. The sample volume should not exceed 10% of the volume of TRIzol used for homogenisation.

   Cells grown in monolayers: Lyse cells directly in a culture dish by adding 1 ml of TRIzol to a 3.5-cm diameter dish, and pipetting the cell lysate up and down several times (see Note 2).

   Cells grown in suspension: Pellet cells by centrifugation. Lyse cells in TRIzol by repetitive pipetting. Use 1 ml of TRIzol per 5–10 × 10^6 of animal, plant or yeast cells, or per 1 × 10^7 bacterial cells (see Note 3).

2. Spin down for 10 min at 12,000 × *g* (4°C) to remove the insoluble material from the homogenate and transfer the supernatant into a new tube.
3. Incubate for 5 min at room temperature (15–30°C) to permit the dissociation of nucleoprotein complexes.
4. Add 0.3 VT of chloroform (where VT is the volume of TRIzol used in the homogenisation step) and mix sample by vortexing for 15 s.
5. Incubate for 3 min at room temperature.
6. Spin for 15 min at no more than 12,000 × *g* (4°C).
7. Transfer the colourless upper aqueous phase into a clean tube (see Note 4).
8. Add 0.5 VT of isopropyl alcohol to precipitate the RNA from the aqueous phase.
9. Incubate the samples:

   For miRNAs: from 2 h to overnight at −20°C.

   For mRNAs: 10–15 min at room temperature.
10. Spin down for 10 min at 12,000 × *g* (4°C).

11. Remove the supernatant. Wash the RNA pellet with 1 VT of 75% ethanol. Mix sample by vortexing.
12. Spin down for 5 min at 7,500 × *g* (4°C).
13. Dry briefly the RNA pellet (air-dry for 5–10 min) (see Note 5).
14. Resuspend the RNA pellet in RNase-free ddH_2O (30–50 μl) by pipetting up and down.
15. Quantify the RNA measuring absorbance at 260 nm. Check the integrity of the extracted RNA on a 0.8% TBE agarose gel (see Note 6).

3.1.2. Total Nucleic Acid Isolation from Plant Tissues with High Levels of Polysaccharides and Polyphenols (10)

1. Grind 2 g of fresh tissue (or the whole fruit) using a prechilled mortar and pestle into a fine powder in liquid nitrogen and transfer into a 15-ml tube.
2. Extract the RNA by the sequential addition of these reagents (the volumes are for a 2 g fresh weight sample):

   1 ml 5 M NaCl

   0.5 ml 10% (w/v) SDS

   1.65 ml 1.95% (w/v) Na_2SO_3

   1.75 ml TBE buffer

   0.1 ml 2-mercaptoethanol
3. Vortex and incubate at 65°C for 5 min.
4. Centrifuge at 1,800 × *g* for 5 min at room temperature.
5. Transfer the supernatant into a new 15-ml tube and add an equal volume of Tris-saturated phenol (pH 8).
6. Mix and centrifuge at 12,000 × *g* for 10 min to achieve phase separation.
7. Transfer the upper aqueous phase into a clean tube.
8. Incubate the upper phase with an equal volume of chloroform–isoamyl alcohol (24:1), mix well by vortexing, and centrifuge at 12,000 × *g* for 10 min.
9. Incubate the upper phase with an equal volume of isopropyl alcohol from 2 h to overnight at –20°C to precipitate sRNA.
10. Spin down for 10 min at 12,000 × *g* (4°C).
11. Remove the supernatant. Wash the RNA pellet with 1 ml of 75% ethanol. Mix sample by vortexing.
12. Spin down for 5 min at 7,500 × *g* (4°C).
13. Dry briefly the RNA pellet (air-dry for 5–10 min) (see Note 5).
14. Resuspend the RNA pellet in 50 μl RNase-free ddH_2O.
15. Quantify the RNA measuring absorbance at 260 nm. Check the integrity of the extracted RNA on a 0.8% TBE agarose gel (see Note 6).

3.2. Separation of RNA Samples by Denaturing 15% PAGE

1. Prepare a 15% denaturing polyacrylamide gel as described in the Subheading 3.
2. Prepare the samples by adding 1 volume of 2× loading buffer to the RNA. Incubate the samples for 5 min at 65°C or 1 min at 90°C to denature the RNA and keep them on ice until the loading of the gel (see Note 7).
3. Place the gels into the tank. Be sure that the short plate faces towards the inner reservoir.
4. Add 1× running buffer (TBE or MOPS) to both reservoirs covering the wells, and rinse out wells with a syringe and a 20-gauge needle (see Note 8).
5. Pre-run the gel in 1× running buffer at 100 V for 30 min (see Note 9).
6. Rinse out wells again with a syringe and a 20-gauge needle.
7. Load the samples (see Note 10).
8. Run the gel around 2 h at 80–100 V until the bromophenol blue reaches about 2 cm at the bottom of the gel (see Note 11).
9. Prise apart the glass plates and stain the gel for 5 min in 10 μg/ml EtBr in 1× running buffer.
10. Take a picture of the gel to check the running and that the RNA samples are equally loaded.
11. Wash the gel in 1× running buffer for 5 min.

3.3. Transferring RNA (see Note 12)

3.3.1. Capillary Blotting (see Note 13)

1. Soak the gel in 10 mM Na-phosphate buffer (pH 7.0) for 10 min and subsequently in 20× SSC for an additional 10 min before blotting.
2. Cut a piece of membrane to the appropriate size to match the size of the gel and soak it in distilled water and then in 20× SSC.
3. Half fill a tray with transfer buffer (TBE or MOPS). Make a platform and cover with three Whatman papers saturated in transfer buffer that are in contact with the transfer buffer in the tray as described in Fig. 1 (see Note 8).
4. Place the treated gel on the platform. Avoid trapping any air bubbles between the gel and the Whatman papers because they block the transfer of nucleic acid to the membrane.
5. Cover the Whatman papers on the platform with clingfilm, but avoid covering the gel (cut the clingfilm around the gel).
6. Place the membrane on top of the gel. Avoid trapping any air bubbles (see Note 14).
7. Cut three pieces of Whatman paper and soak them in 20× SSC.
8. Place the soaked Whatman papers on top of the membrane. Avoid trapping any air bubbles.

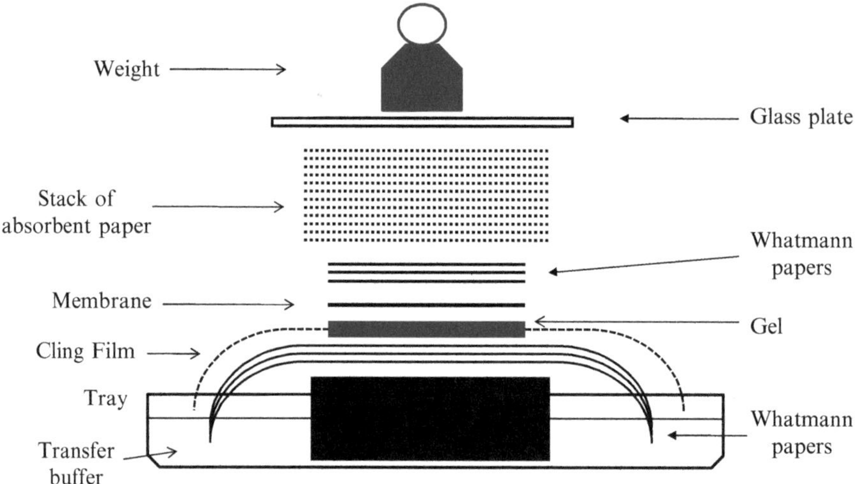

Fig. 1. A typical arrangement for capillary blotting.

9. Place a stack of absorbent paper towels at least 5 cm high on top of the Whatman papers.
10. Place a glass plate and a weight over the paper stack. The weight should not exceed 750 g for a 20 × 20-cm gel.
11. Let transfer proceed overnight.
12. After blotting, carefully dismantle the blotting system and label the membrane.

3.3.2. Semidry Blotting

1. Cut out six rectangles of Whatman paper and one piece of membrane of the same size as the gel with clean gloves and a razor.
2. Soak them in 1× transfer buffer (TBE or MOPS) (see Note 8).
3. On a semidry blotter, place three Whatman papers (on top of each other in a stack). Roll over a 5-ml tip to remove bubbles.
4. Place the membrane and the gel on top (see Note 14).
5. Place the three remaining soaked Whatman papers. Remove air bubbles by rolling over a 5-ml tip.
6. Transfer for 35 min at 3 mA/cm^2 (190 mA/mini gel) in the cold room.
7. Disassemble, label blot, and rinse in 1× transfer buffer.

3.4. Cross-linking

3.4.1. UV Cross-linking

1. Place membrane, RNA side up, on a dry piece of Whatman paper for 5 min at room temperature.
2. UV cross-link by using an optimised protocol or, alternatively, illuminate for up to 2 min with a bench-top UV lamp (see Notes 15 and 16).

3.4.2. Chemical Cross-linking

1. Place a Whatman paper on a large piece of Saran wrap.
2. Add the cross-linking solution to the Whatman paper (5 ml is enough).
3. Place the membrane on top of the Whatman paper with the RNA side up.
4. Wrap the Saran wrap around to make a sealed parcel.
5. Incubate at 60°C for 1–2 h.
6. Wash the membrane in ddH_2O for 10 min on a shaker.
7. Wash again in ddH_2O for 10 min on a shaker.
8. Wrap in Saran wrap and store in the fridge. This membrane can be stored for several months under these conditions.

3.5. Hybridisation

3.5.1. Pre-hybridisation

1. Preheat a hybridisation oven to 37°C.
2. Prewarm the hybridisation buffer (ULTRAhyb-oligo hybridisation buffer from Ambion) at 37°C to redissolve any precipitated material.
3. Place the cross-linked membrane into a hybridisation bottle and add the prewarmed hybridisation buffer (1 ml/10 cm^2 of membrane). Make sure that enough solution is present to keep the membrane uniformly wet (see Note 17).
4. Rotate the bottle horizontally in the hybridisation oven from 2 h to overnight.

3.5.2. Probe Preparation

1. Prepare the probe in a screw-cap Eppendorf (VT = 20 μl).

10 μM oligonucleotide	2 μl
5× polynucleotide kinase (PNK) forward reaction buffer	4 μl
ddH_2O	11 μl
γ-^{32}P-ATP	2 μl
T4 PNK from Invitrogen	1 μl

The oligonucleotide should be reversed complementary to the short RNA to be detected.

2. Incubate for 1 h at 37°C.
3. Snap cool on ice and add 20 μl of sterile water.

3.5.3. G25 Sephadex Column Preparation (see Note 18)

1. From the tube opening, puncture the bottom of a sterile, nuclease-free, 0.5-ml microcentrifuge tube 1–2 times with a 21-gauge needle.
2. Place the 0.5-ml microcentrifuge tube into a sterile 2-ml microcentrifuge tube.
3. Add 10 μl of glass beads to the 0.5-ml tube.

4. Add 0.5 ml of G25 Sephadex matrix and centrifuge for 2 min at 1,800 × *g*.
5. Discard the eluate and centrifuge for another 2 min at 1,800 × *g*.

3.5.4. Hybridisation

1. Purify the probe through the G25 Sephadex column to remove unincorporated nucleotides for 2 min at 1,800 × *g*.
2. Add the probe to the hybridisation tube (see Note 19).
3. Hybridise at 37°C overnight by rotating the tube in the hybridisation oven (Note 20).

3.5.5. Wash and Exposure

1. Membranes are washed twice for 30 min each at 37°C with about 20 ml of wash buffer. Keep rotating the tube between washes (see Notes 21 and 22).
2. Remove the membrane from the hybridisation bottle and wrap it in Saran wrap. Remove air bubbles between the membrane and the Saran wrap.
3. Expose the membrane to X-ray film or phosphorimager plate (see Note 23).
4. Develop film or scan phosphorimager plate (see Note 24).

3.5.6. Stripping

1. Heat the stripping solution at 100°C.
2. Soak the membrane for 20 min and shake gently.
3. Check the removal of the probe using the appropriate detection system.
4. Repeat if necessary.
5. Seal and expose for as long as you anticipate exposing your next probe.

4. Notes

1. Pre-chill the mortar, pestle, and tubes before starting to grind the sample. Always keep the samples frozen adding N_2 (l) until TRIzol is added. TRIzol maintains the integrity of the RNA, while disrupting cells and dissolving cell components.
2. The amount of TRIzol added is based on the area of the culture dish (1 ml/10 cm^2) and not on the number of cells present. An insufficient amount of TRIzol can result in the contamination of the isolated RNA with DNA.
3. Avoid washing cells before the addition of TRIzol because this increases the possibility of mRNA degradation. The disruption of some yeast and bacterial cells can require the use of a homogeniser.

4. Save the organic phase if you are interested in isolating DNA or protein from your sample.
5. It is important not to let the RNA pellet dry completely because this will decrease its solubility.
6. To prepare the sample, add 5 μl of loading buffer to 5 μl of the RNA sample and incubate at 65°C for 5 min or at 95°C for 1 min. Keep them on ice until the loading of the gel. The ribosomal RNA bands should be visible and show no substantial smear.
7. For sRNA detection, load 10–15 μg of total RNA per lane. If the sample is not concentrated enough use a speed-vac. If it is difficult to detect a certain short RNA, the amount of RNA can be increased up to 40–50 μg, although the running can be affected because less RNA generates a better resolution and clear bands. If you need to load a larger amount of RNA, it is better to use thicker gels (1.0 mm instead of the standard 0.75 mm). Sensitivity can also be increased by one of the following modifications: using the chemical cross-linking method instead of the UV; using a LNA primer as a probe (14); or purifying and loading the sRNA fraction. In the latter, comparison among samples is not allowed. The recommended amount of sRNA-enriched samples is 4 μg per well. We normally run 0.75-mm 15% gels and use semidry blotting. These conditions are usually optimal to separate 10–15 μg of total RNA.
8. When using chemical cross-linking, it is recommended to use MOPS–NaOH buffer instead of the conventional TBE because Tris has primary amine groups that can react with EDC.
9. This pre-run means that the gel will be warmed up, which helps to keep the RNA denaturated during electrophoresis, resulting in a better resolution. Pre-running is especially recommended for larger gels.
10. Load a DNA primer, with the same sequence as the sRNA of interest, especially if it is the first time this blot has been done. This DNA works as a positive control for labelling, hybridisation, and detection. Load it into a separate lane and keep an empty lane between the primer and samples to avoid contamination.
11. In these conditions, bromophenol blue runs around 15 nt and xylene cyanol around 60 nt. The sRNA fraction is found in between.
12. The type of membrane used for transfer depends on the cross-linking method applied. If using the UV cross-linking method, either positively charged (such as Hybond N+; Amersham) or neutral (such as Hybond NX; Amersham) membranes can be used. Positively charged membranes have

a higher capacity; therefore, more RNA can be bound to the membrane. This can help achieve a stronger signal. However, it can also be a disadvantage if it leads to a higher background owing to the unspecific binding of the probe. When using the chemical cross-linking method, positively charged membranes cannot be used because the RNA binds only to neutral membranes under these conditions.

13. Capillary blotting is less efficient for thicker gels. Moreover, decreasing the gel concentration to 8 or 10% is recommended. In this case, you can run a longer gel than the Bio-Rad Mini Protean II.
14. Mark the gel and the membrane by cutting the same corner to establish the orientation in the membrane according to the gel.
15. Make sure that the RNA-carrying side of the membrane faces towards the UV source.
16. The optimum exposure for Hybond NX is 70,000 $\mu J/cm^2$. To optimise the cross-linking procedure, prepare several identical control blots on the membrane of choice. Cover the transilluminator with Saran wrap to protect the surface of the membrane. Expose each blot for a different length of time (15 s to 5 min). Hybridise all blots in the same hybridisation tube with an appropriate probe. The optimum UV exposure time will be indicated by selecting the blot showing the strongest signal.
17. Remove bubbles between the membrane and hybridisation tube using long forceps. Make sure that the membrane unwinds in the bottle and is not creased because this can cause areas of high background.
18. To minimise background, remove unincorporated γ-^{32}P-ATP from the labelled probe before hybridisation.
19. It is important that the undiluted probe solution does not touch the membrane, because this can generate spots on the membrane. Add the probe directly into the hybridisation buffer or pour the pre-hybridisation buffer into a tube, add the probe, mix well, and then immediately pour back into the hybridisation tube containing the membrane. In case of a high background, the pre-hybridisation buffer can be discarded and the probe should be added into fresh hybridisation buffer. In this case, it is important to heat up the new batch of hybridisation buffer to the hybridisation temperature.
20. If an LNA probe is used, the time of hybridisation can be reduced to as little as 4 h. The hybridisation temperature can be increased up to 50°C and more stringent washing conditions can be applied.

21. Washing together membranes hybridised with different probes is not recommended to avoid cross-hybridisation.
22. It is important that membranes are not completely dried after the final wash. Dried membranes cannot be stripped.
23. Suggested initial exposure times are 4 h for abundant sRNA and from overnight to a few days for non-abundant sRNAs.
24. Equal loading is checked by re-probing the membrane with U6 specific primer. The primer we use for U6 detection is 5′GCTAATCTTCTCTGTATCGTTCC 3′. The U6 signal is about one-third from the top of the gel, whereas the short RNA signal is usually about two-thirds from the top.

Acknowledgments

The author wishes to thank Dr. Alvarez for reading the manuscript. S.L-G is supported by a Spanish Government postdoctoral fellowship.

References

1. Phillips, J. R., Dalmay, T., and Bartels, D. (2007) The role of small RNAs in abiotic stress, *FEBS Lett* 581, 3592–3597.
2. Lee, Y., Ahn, C., Han, J., Choi, H., Kim, J., Yim, J., Lee, J., Provost, P., Radmark, O., Kim, S., and Kim, V. N. (2003) The nuclear RNase III Drosha initiates microRNA processing, *Nature* 425, 415–419.
3. Tomari, Y., and Zamore, P. D. (2005) Perspective: machines for RNAi, *Genes Dev* 19, 517–529.
4. Reinhart, B. J., Weinstein, E. G., Rhoades, M. W., Bartel, B., and Bartel, D. P. (2002) MicroRNAs in plants, *Genes Dev* 16, 1616–1626.
5. Peragine, A., Yoshikawa, M., Wu, G., Albrecht, H. L., and Poethig, R. S. (2004) SGS3 and SGS2/SDE1/RDR6 are required for juvenile development and the production of trans-acting siRNAs in Arabidopsis, *Genes Dev* 18, 2368–2379.
6. Vazquez, F., Vaucheret, H., Rajagopalan, R., Lepers, C., Gasciolli, V., Mallory, A. C., Hilbert, J. L., Bartel, D. P., and Crete, P. (2004) Endogenous trans-acting siRNAs regulate the accumulation of Arabidopsis mRNAs, *Mol Cell* 16, 69–79.
7. Lu, C., Kulkarni, K., Souret, F. F., MuthuValliappan, R., Tej, S. S., Poethig, R. S., Henderson, I. R., Jacobsen, S. E., Wang, W., Green, P. J., and Meyers, B. C. (2006) MicroRNAs and other small RNAs enriched in the Arabidopsis RNA-dependent RNA polymerase-2 mutant, *Genome Res* 16, 1276–1288.
8. Borsani, O., Zhu, J., Verslues, P. E., Sunkar, R., and Zhu, J. K. (2005) Endogenous siRNAs derived from a pair of natural cis-antisense transcripts regulate salt tolerance in Arabidopsis, *Cell* 123, 1279–1291.
9. Chomczynski, P., and Sacchi, N. (1987) Single-step method of RNA isolation by acid guanidinium thiocyanate-phenol-chloroform extraction, *Anal Biochem* 162, 156–159.
10. Kumar, G. N., Iyer, S., and Knowles, N. R. (2007) Extraction of RNA from fresh, frozen, and lyophilized tuber and root tissues, *J Agric Food Chem* 55, 1674–1678.
11. Southern, E. M. (1975) Detection of specific sequences among DNA fragments separated by gel electrophoresis, *J Mol Biol* 98, 503–517.
12. Pall, G. S., Codony-Servat, C., Byrne, J., Ritchie, L., and Hamilton, A. (2007) Carbodiimide-mediated cross-linking of RNA to nylon membranes improves the detection of

siRNA, miRNA and piRNA by Northern blot, *Nucleic Acids Res* 35, e60.

13. Vester, B., and Wengel, J. (2004) LNA (locked nucleic acid): high-affinity targeting of complementary RNA and DNA, *Biochemistry* 43, 13233–13241.

14. Valoczi, A., Hornyik, C., Varga, N., Burgyan, J., Kauppinen, S., and Havelda, Z. (2004) Sensitive and specific detection of microRNAs by Northern blot analysis using LNA-modified oligonucleotide probes, *Nucleic Acids Res* 32, e175.

Chapter 4

Profiling MicroRNAs by Real-Time PCR

Nana Jacobsen, Ditte Andreasen, and Peter Mouritzen

Abstract

A variety of physiological processes are associated with changes in microRNA (miRNA) expression. Analysis of miRNA has been applied to study normal physiology as well as diseased states including cancer. One major challenge in miRNA research is to accurately and practically determine the expression level of miRNAs in various experimental systems. Many genome-wide miRNA expression profiling studies have relied on microarrays technology, and frequently differentially expressed miRNAs have subsequently been confirmed with real-time quantitative PCR studies. Here, we describe how different primer strategies for first-strand cDNA synthesis and PCR amplification can affect measurements of miRNA expression levels. Overcoming the small nature of miRNAs is a difficult task as the short sequence available does not allow for designing primers using standard PCR primer design guidelines. Finally, we demonstrate how to determine differentially expressed miRNAs using a locked nucleic acid-based real-time PCR approach.

Key words: Quantitative real-time PCR, Reverse transcription, Amplification efficiency, Quantification cycle, SYBR Green I detection, Primer design strategies, Melting temperature

1. Introduction

The number of known non-protein coding small RNAs has increased considerably during the last decade. Amongst the different classes of small RNAs, microRNAs (miRNAs) stand out as the best characterized class in terms of numbers, diversity, mode of expression, and function (1). The miRNAs are generated from hairpin structures maturated by the RNAse III ribonucleases Drosha/Dicer to mature miRNAs of ~22 nucleotides (nt). The mature miRNAs act as post-transcriptional regulators of gene expression by base pairing with messenger RNAs (mRNAs), thereby causing exonucleolytic mRNA decay or translational repression. Unquestionably, small RNAs work as essential modulators in the

Tamas Dalmay (ed.), *MicroRNAs in Development: Methods and Protocols*, Methods in Molecular Biology, vol. 732, DOI 10.1007/978-1-61779-083-6_4, © Springer Science+Business Media, LLC 2011

immensely complicated framework that regulates the transcription of the human genome (1).

Quantification of miRNAs by real-time quantitative PCR (qPCR) represents a challenge impeded by the short nature of the miRNAs (19–25 nt) and the high variability in the %GC nucleotide content. Also, different miRNA families of homologous closely related sequences exist within a given organism. The technical dilemma of performing qPCR on the short miRNAs is due will the fact that it is impossible to design the required two PCR primers within the available 19–25 nt for a specific PCR . In general, a single DNA primer is 18–24 nt long to achieve the optimal specificity and annealing temperature toward the target (2) about the same length as a miRNA. Various reverse transcription (RT) qPCR techniques have been developed (3, 4). Common to all techniques, a non-target-related nucleotide tag is added at the 3′-end of miRNA during the conversion of the miRNA into cDNA. On the extended cDNA, there is now room for a new primer-binding site together with the miRNA-specific primer-binding site. For obvious reasons, the added nucleotide tag can be designed to attain a rational PCR primer sequence; the real challenge is often designing the miRNA-specific PCR primer. Here low %GC content can make it difficult to design a PCR primer with high enough annealing temperature and vice versa, i.e. if the %GC content is very high, primers will tend to have too high annealing temperature. The low %GC content miRNAs require longer PCR primers and may eventually lead to a full-length miRNA primer. A full-length miRNA primer often tends to be associated with severe PCR contamination problems and thereby lead to frequent misinterpretations of the results.

Likewise, short nucleotide stretches of palindromic sequence within the miRNA may be difficult to avoid. Primers designed including such palindromic regions may result in amplification of unspecific primer-dimers. Also, primer-dimers may arise from the 3′-end complementarity of the primer pair. The close resemblance of miRNAs sequence belonging to the same family makes the discrimination of these difficult, and requires considerable amounts of time spent on the assay design or optimization steps.

We have found that using the high affinity nucleotide analog locked nucleic acid (LNA) (5) provides a very useful flexibility in primer design. The annealing temperature of the primers can be increased by exchanging DNA nucleotides with LNA within the primer. Concurrently, the optimal annealing temperature of the primer can be maintained even though the nucleotide length of primers is shortened. Designing shorter PCR primers spiked with LNA gives a higher degree of design freedom to avoid palindromic sequence or primers that are complementary to each others. Finally, LNA increases the ability of primers to discriminate between sequences with close resemblance, thus facilitating discrimination of nearly homologous miRNAs within the same family.

1.1. Strategies in Quantitative Real-Time RT-PCR of miRNA

1.1.1. Reverse Transcription and Primer Tag Addition

The optimal reverse transcription is a universal reaction unbiased towards particular sequences. Significant variation is introduced in the reverse transcription step (6) and, therefore, the ideal reverse transcription reaction will generate cDNA for all the different RNA molecules to be analyzed by real-time RT-qPCR and will keep the amount of variation introduced independent of the different molecules to be analyzed. Random hexamer priming is close to fulfill the universality requirements and is commonly used in a classical real-time RT-qPCR on mRNA. However, miRNAs are short and the stoichiometry of the random hexamer priming toward the miRNAs is altered due to the fact that the number of primer annealing sites is much less than the primer-binding sites to larger RNAs (e.g., mRNAs) (7). One side effect is that many truncated cDNA molecules will be generated and this is unacceptable considering the short length of miRNAs. The miRNA-specific methods developed have in common that a nucleotide tag is added to the 3′-end of the miRNA and thereby introduce a non-miRNA-specific primer-binding site for a PCR primer. Two of the well-described methods are (1) polyadenylation of the miRNA at 3′-end followed by poly-dT primed reverse transcription, and (2) miRNA-specific reverse transcription where a gene-specific primer is designed toward the 3′-nucleotides of the miRNA in question. Truly, the polyadenylation reaction is the most universal method, but in our hands, the required specificity and sensitivity of the entire assay are obtained only when combined with the use of high-affinity DNA analogs such as LNA in the primers. While the gene-specific reverse transcription provides high levels of specificity and sensitivity, the only drawback is the lack of universality, as one first strand cDNA synthesis reaction must be performed per analyzed miRNA. Universality may be achieved by multiplexing the reverse transcription primers, but this approach requires careful optimization and would seem prone to bias or skewing when ratios of miRNAs change relative to each other.

1.1.2. Detection Methods

The two most common detection methods in real-time qPCR are intercalating dyes (e.g. SYBR Green I) (8) and fluorescent probes (e.g. 5′-hydrolysis probes) (9). Both methods intend to measure in real time the increasing number of a specific amplified sequence. The advantage of intercalating dye-based methods is that it allows a specificity post-analysis (melting curve analysis) at the end of the PCR to verify that only one amplicon with the expected/predicted melting temperature (T_M) and no primer-dimers have been generated in the PCR. The melting curve analysis involves denaturing of all double-stranded DNA molecules, including amplicons and primer-dimers, followed by a stepwise lowering of the temperature, where the binding of SYBR Green I is measured while the amplicon re-anneals (8). The 5′-hydrolysis probe-based methods rely on a different strategy to provide specificity of the PCR. End-point analysis of amplicons and primer-dimers is not an option; rather, the 5′-hydrolysis probe intends to provide specificity by targeting a third recognition site (apart from the two primer sites) in the amplicon (9). Only amplicons that have this third recognition site will result in fluorescence increase through hydrolysis of the probe, whereas other amplicons will not be measured. Probe hydrolysis results from the 5′-3′ exonuclease activity of *Taq* polymerase cleaving off a fluorescent quenching moiety from the probe when it is hybridized to a complementary target (9). As long as primers are specific leading to generation of one amplicon only and no primer-dimers, the efficiency of the 5′-hydrolysis probe-based methods will be good. The assay efficiency will be affected if more amplicons and/or primer-dimers are formed in which case the specificity of the probe becomes crucial. In a classical real-time RT PCR on mRNA (protein coding RNAs), where a 60–150-bp amplicon is generated, the specificity provided by 5′-hydrolysis probes is good (10). Here, the available sequence length is long enough to span two primer-binding sites and a unique probe-binding site. However, positioning of the probe is really difficult when it comes to shorter targets like miRNAs. The difficulty arises when trying to find a target position that has not already been used for primer binding either by the RT primer or the PCR primers. A possible position for the probe can be between the forward primer and the reverse primer or between the forward primer and the reverse transcription RT primer, overlapping at least two different primer sites. Often the forward primer and the reverse transcription primer overlap and, therefore, the probe will not provide specificity toward miRNA as such. If the position of the probe is annealing to one primer only, the detection probe will not be able to distinguish between primer-dimers and the amplicon of interest.

1.1.3. Mature and Precursor Forms of miRNAs

The mature forms of miRNAs are generated from miRNA precursors that are hairpin structures called pre-miRNAs (7). Since the mature miRNA is believed to be the biological active molecule, there is a interest from researches to detect this form selectively. Selective purification of one or the other form of miRNAs is not an option during RNA isolation or subsequent purifications and, therefore, the detection method should ideally facilitate discrimination between mature and precursor forms. The mature miRNA may originate from both the 3′- and the 5′-arm of the pre-miRNA hairpin molecule. Universal to all detection technologies including array-based methods is that discrimination between the mature and the precursor form is most successful when the miRNA is located on the 5′-arm of the pre-miRNA hairpin molecule. Discrimination between the mature and the precursor form is less successful when the mature miRNA is located on the 3′-arm of the pre-miRNA hairpin molecule. In RT-qPCR techniques, different reverse transcription strategies have been developed to improve the discrimination of mature miRNA and precursor forms. The different strategies have evolved around the design of the reverse transcription primer resulting in designs where the primer either is linear or contains a stem-loop. The linear reverse transcription primer is used both in gene-specific reverse transcription of miRNAs and poly-dT reverse transcription of polyadenylated miRNAs. It has been shown that if the mature miRNA originates from the 5′-arm of the precursor molecule, the linear reverse transcription primers provide more than 2,000-fold discrimination between the pre-miRNA and the mature miRNA (3). This high level of discrimination is probably caused by the pre-miRNA hairpin loop making steric hindrance toward the RT primer. The stem-loop reverse transcription primer consists of a miR-specific part and a tagging sequence forming a hairpin loop and a small stem (9). The base stacking of the stem may enhance the thermal stability of the RNA–DNA heteroduplex and, furthermore, spatial constraint of the stem-loop is thought to improve the assay specificity toward mature miRNA. However, the stem-loop RT-primer approach has more difficulty in discriminating between the mature miRNA originating from the 3′-arm of the precursor and precursor molecules themselves because of the similarities of the 3′-termini of the two miRNA forms. Interestingly, we have found a similar pattern in the ability to discriminate between mature and precursor miRNAs for microarrays that rely on either linear or stem-loop capture probes (data not shown). Though stem-loop capture probes in theory should have a preference for mature miRNAs, these are particularly difficult to discriminate from precursors when the mature miRNA originates from the 3′-arm. Therefore, we have concluded that different detection methods will be needed if improved discrimination is required for mature miRNAs and their precursors.

1.1.4. The 3′-Heterogeneity of miRNAs

Recently, it has been discovered that many of the mature miRNAs exhibit an extensive degree of polymorphisms, primarily at 3′-ends and to a lesser extent at the 5′-end (11). Of these variants, the miRNA annotated in Sanger's microRNA registry miRBase (12) is found to be the most abundant. At this point, little is known about the biological significance of the 3′-heterogeneity, but since the miRNA seed sequence is not affected by this heterogeneity, it may be speculated that the 3′-end miRNA variants are biological active. Therefore, it may be argued that the detection technologies should allow for the detection of more of these 3′-end variants in order to make a reasonable assessment of the biologically active molecules. We have found that the heterogeneity of the mature miRNA 3′-end affects the detection efficiency of stem-loop reverse transcription RT primer-based real-time PCR. Since the exact match of 3′-terminal of miRNA is the basis for high efficiency of this method, the 3′-end variations negatively affect detection by this method. One-nucleotide deletion of the miRNA decreased the efficacy of stem-loop-based real-time PCR by two quantification cycles, which translates into a 25% efficiency of detection of the truncated miRNA (11), compared to that with the normal miRNA annotated in miRBase. Conversely, we have found that linear reverse transcription primer-based real-time PCR detects 3′-end variants with almost equal efficiency as the normal full-length miRNA found in miRBase (12).

1.2. Experimental Design Guidelines

A basic experimental design compares differences between a set of biological samples (i.e. treated vs. untreated, sick vs. normal, or it compares developmental stages, or follows time courses). One of the most common real-time RT-qPCR experimental designs is a nested design where the qPCR samples are nested in the RT reactions which again are nested in the biological samples. For each biological sample group, one or more subject(s) could be included. For the RT and qPCRs, technical repeats are usually included. In general, the following experimental set-up is recommended for each biological replicate: two RT reactions followed by two PCRs for each RT reaction, giving rise to four technical data points for each miRNA and sample. However, since usually the RT reaction gives rise to most of the variations, it may be advisable to perform three RT reactions with 1–2 PCRs for each sample, if sample amount and economy allow. It is further recommended to always include at least three biological replicates of each sample type in order to allow statistical analysis of the results. If small changes in miRNA expression are expected, it may be necessary to include more biological replicates to ensure a significant result. In general, it is recommended that replicates should be included at any stage during sample procurement, processing, RNA isolation, etc., which could give rise to variations between samples. When the measurements are independent, e.g., individual gene-specific RT reactions and individual qPCR assays, technical repeats might be averaged before data normalization.

The purpose of normalization is to remove technical and biological variations between samples that are not related to the biological changes under investigation. Proper normalization is critical for the correct analysis and interpretation of results from real-time PCR experiments (6).

In general, it is recommended to employ as many controls (reference genes) as possible to guarantee proper real-time PCR quantification. The controls should be empirically validated for each study to identify the 2–3 controls that are stably expressed across the panel of studied samples. Different programs exist for evaluating the best performing endogenous controls (e.g., GeNorm (13), NormFinder (14), and BestKeeper (15)) and for applying multiple endogenous controls for normalizing target expression (e.g. qBASE (13) and REST (16)). Since the method(s) of choice for normalizing real-time PCR data are highly individual, it is recommended to visit the following Web site for detailed information about the approaches, methods, and software available for real-time PCR quantification: http://www.gene-quantification.info/.

One simplified approach for normalizing target expression with the miRCURY LNA™ endogenous controls is to apply the comparative quantification "$\Delta\Delta C_T$" method, also known as the $2^{-\Delta\Delta C_T}$ method. This method relies on comparing the differences in quantification cycles (Cq) values obtained for the targets(s) and selected control(s) in a sample of interest with the Cq values obtained for the same targets(s) and control(s) in a reference sample (e.g., tumor vs. normal tissue). When using this approach, it is critical that the selected controls are not regulated by the experimental conditions. Hence, in order to normalize miRNA expression levels, real-time PCR is performed with the parallel use of the miRNA primer set(s) of interest and the endogenous control primer sets (e.g., miRCURY LNA™ endogenous control primer sets).

Finally, the importance of including a proper set of technical controls should be emphasized. It is recommended to include both reverse transcription controls and non-template controls for each miRNA assay performed. The reverse transcription control is performed by omitting reverse transcriptase in the first-strand cDNA synthesis reaction. If a PCR product is amplified from this control reaction, it indicates genomic DNA contamination of the RNA template. The non-template controls are conducted at the real-time PCR amplification step by omitting cDNA template in the real-time PCRs. This control will reveal PCR product contamination of the reaction. Also, the aforementioned controls will give an indication of the natural level of background of a particular assay with the given laboratory setup defining the detection limit threshold.

1.3. Sample Preparation

The importance of sample preparation is often overlooked, but in order to find biological relevant differences between sample groups, it is important to consider whether the samples have undergone the same handling procedures. Therefore, standardization and systematic treatment of the material are needed, including a description of the origin of the sample (tissue section or cell culture), whether it was micro-dissected or macro-dissected, how the sample storage conditions and duration were (fresh, frozen, formalin-fixed, or paraffin-embedded), and which method of nucleic acid extraction was employed.

Several methods for extraction and purification of RNA suited for a RT-qPCR experiment have been described. A frequently applied method for total RNA purification is acid guanidinium thiocyanate–phenol–chloroform extraction followed by ethanol precipitation and centrifugation (17). Another method is the silica-based affinity extraction (18). Many manufactures offer kits for RNA purification. In general, the columns in these kits have a nucleotide cutoff above 200 nt. For successful extraction and purification of small-sized RNAs it is, however, necessary to apply a suitable affinity matrix and a modified protocol.

The purified RNA has to be validated with respect to quantity and quality (19). This is conveniently obtained by spectrophotometry, gel electrophoresis, microfluidity, or fluorescent dye detection. The purity of the RNA sample can be validated by measurement of the A_{260}/A_{280} and A_{260}/A_{230} ratios which are sensitive to the presence of residual phenol and guanidinium thiocyanate. The advantage of using gel electrophoresis/microfluidic methods is that they provide both a quantitative measurement and a quality evaluation of the general RNA purification conditions such as RNA degradation and genomic DNA contamination.

1.4. Working with RNA

RNA is not protected during the preparation of tissue. Thus, it is important that the procedure is carried out as quickly as possible, particularly at the cell lysate preparation step.

Fresh or frozen tissues may be used for the procedure. Tissues should be flash-frozen in liquid nitrogen and transferred immediately to a –70°C freezer for long-term storage. Tissues may be stored at –70°C for several months. Thawing during excising and weighing prior to RNA preparation can reduce the integrity of the RNA preparation.

The purified RNA sample may be stored at –20°C for a few days. It is recommended that samples are stored in siliconized tubes and placed at –70°C for long-term storage.

Working with RNA is an ongoing battle against RNases. The RNases are very stable and robust RNA degrading enzymes. Autoclaving of solutions and glassware is not always sufficient to completely inactivate these enzymes. It is recommended as a first step to create an RNase-free environment following the precautions below.

The RNA area should be located away from microbiological work stations.

Clean, disposable gloves should be worn at all times when handling reagents, samples, pipettes, disposable tubes, etc. Also, it is recommended that gloves are changed frequently to avoid contamination.

Designated solutions, (filter)tips, tubes, laboratory coats, pipettes, etc., for RNA work only.

All RNA solutions should be prepared using at least 0.05% DEPC-treated autoclaved water or molecular biology grade nuclease-free water.

Clean all surfaces with commercially available RNase decontamination solutions.

When working with purified RNA samples, ensure that they remain on ice.

1.5. Experimental Preparations

Here and in the following sections of this chapter, we describe the critical steps in conducting a real-time RT-qPCR experiment using the miRCURY LNA™ microRNA PCR System as an example.

The miRCURY LNA™ microRNA PCR System has been replaced with the miRCURY LNA™ Universal RT microRNA PCR, which is a performance-wise improved technology with increased sensitivity and specificity. Furthermore, the protocol has been simplified to allow a one-tube universal cDNA reaction for all miRNAs. However, the overall principles and advantages of LNA apply as does the guidelines for normalization and data analysis.

The real-time RT-qPCR detection of miRNA expression profiling is based on the miRCURY LNA™ microRNA PCR System. The system comprises a first-strand cDNA synthesis kit, a SYBR® Green master mix, and miRNA and endogenous control primer sets. The detection is performed in total RNA samples according to a two-step protocol and is based on SYBR Green I detection.

Prior to starting the real-time RT-PCR experiment, place the 5× RT reaction buffer, miR-specific or the control-specific primer sets, dNTP mix, and nuclease-free water on ice and thaw for 15–20 min. Immediately before use, remove the enzymes (reverse transcriptase, RNase inhibitor, and SYBR® Green master mix) from the freezer, mix by flicking the tubes, and place on ice. Protect the SYBR® Green master mix vials from light. The rest of the reagents are mixed by vortexing.

It is recommended to prepare an RT master mix of the relevant reagents and a PCR master mix including the primers when performing first-strand cDNA synthesis on multiple RNA samples, and include 10% excess of all reagents in order to allow for pipetting losses.

1.5.1. First-Strand cDNA Synthesis

The miRNA-specific first-strand cDNA synthesis kit converts the miRNA by reverse transcription (RT) into a cDNA template in the first step using a miR-specific RT primer for the miRNA of interest. The use of a miR-specific primer and the thermostable and sensitive Reverse Transcriptase ensures a highly specific cDNA synthesis reaction. The cDNA template can be generated directly from a total RNA sample. Prior miRNA enrichment is not needed. However, the method used for purification of total RNA must preserve small RNAs.

1.5.2. Real-Time PCR Amplification

The miRNA-specific real-time PCR is performed in the second step with the cDNA template from the first-strand cDNA synthesis reaction. A miRNA-specific primer tags the 5′-end of the miRNA and a universal PCR primer tags the RT primer during this PCR amplification. The design of the short LNA-enhanced primer enables a miRNA sequence-specific amplification by direct annealing to miRNA primer-binding sites, and a highly sensitive and accurate miRNA real-time PCR is achieved when this is combined with the use of SYBR Green I detection, which allows monitoring fluorescence at 520 nm.

2. Materials

1. The miRCURY LNA™ First-strand cDNA synthesis kit, 200 reactions (Exiqon, Denmark).
2. The miRCURY LNA™ SYBR® Green master mix, 200 reactions including the universal primer (Exiqon, Denmark).
3. A selection of miRCURY LNA™ miRNA primer sets for 50 RT reactions and 100 PCRs (Exiqon, Denmark).
4. A selection of miRCURY LNA™ control primer sets for 50 RT reactions and 100 PCRs (Exiqon, Denmark).
5. Nuclease-free low nucleic acid-binding micro-centrifuge tubes, and PCR tubes/plates.
6. Nuclease-free filter barrier pipette tips.
7. Heating block, thermal cycler, or other incubators suited for temperatures of 50–90°C.
8. A real-time PCR instrument (e.g. Roche LightCycler 480 or ABI 7500 system).
9. Optional, ROX passive reference dye (Invitrogen, USA).
10. Optional, Uracil-DNA Glycosylase (UDG) (Invitrogen, USA).

3. Methods

3.1. The First-Strand cDNA Synthesis Protocol

1. Adjust each of the template RNA samples to a concentration of 1–10 ng in 4.5 µl using nuclease-free water (see Notes 1 and 2).
2. For each sample, prepare a duplicate of the RT master mix containing 2 µl 5× RT reaction buffer, 2 µl miR-specific or control-specific RT primer (Vial A, red), 0.5 µl dNTP mix, 0.5 µl RNase inhibitor, and 0.5 µl reverse transcriptase, and place on ice. The no-RT enzyme control is prepared likewise, but with nuclease-free water instead of reverse transcriptase (see Notes 3 and 4). Gently mix and spin the RT master mix.
3. Dispense 5.5 µl of RT master mix into nuclease-free tubes and place 4.5 µl of template RNA in each tube. It is important to keep reagents and reactions on ice (or at 4°C) at all times. Mix the reaction by gentle vortexing or pipetting to ensure that all reagents are mixed thoroughly. Spin down after mixing (see Note 5).
4. Transfer the tubes to a heating block or a PCR instrument and incubate for 30 min at 50°C followed by heat inactivation of the reverse transcriptase for 5–10 min at 85°C. Although not recommended, the protocol can be paused at this stage. The cDNA may be left overnight at 4°C or for longer time at –20°C.

3.2. The Real-Time PCR Amplification Protocol

1. Dilute the cDNA reactions 1:10 in nuclease-free water to allow pipetting of a larger volume of the reaction into the subsequent real-time PCR, as this generally results in more accurate pipetting and reproducible results.
2. When multiple real-time PCRs are performed, it is recommended to prepare a master mix of all other reagents except the cDNA template. Prepare the PCR master mix including a 10% excess of all reagents. For the miR-specific primer sets, prepare two times the PCR master mix for each sample and an additional reaction for the NTC control containing 10 μl 2× SYBR® Green master mix, 1 μl LNA™ PCR primer (Vial B, blue), 1 μl Universal PCR primer (Vial C, blue), and 4 μl nuclease-free water, and place it on ice. Equally, for the endogenous control primer sets, prepare two times the PCR master mix for each sample and an additional reaction for the NTC reaction containing 10 μl 2× SYBR® Green master mix, 2 μl control PCR primer mix (Vial B, blue), and 4 μl nuclease-free water (see Notes 6 and 7). Gently mix and spin the reagents.
3. Dispense 16 μl of PCR master mix in PCR tubes or wells of a PCR plate and add 4 μl diluted cDNA template to each tube except for the NTC reactions. Instead 4 μl of nuclease-free water should be added to the NTC reactions. Cap the tubes or seal the plate with sealing tape suited for fluorescent detection. Mix the reactions by gentle vortexing to ensure that all reagents are mixed thoroughly, and spin down to ensure that no droplets on the walls and air bubbles are left in the reaction volume.
4. On a Roche LightCycler 480 (LC480), perform 40 cycles of PCR amplification followed by melting curve analysis using the following settings: polymerase activation and denaturing for 10 min at 95°C, 40 amplification cycles consisting of 10 s at 95°C and 20 s at 60°C with optical reads conducted at the 60°C step, and finally melting curve analysis according to the instrument manufacture's recommendation.

3.3. Raw Data and Data Analysis

1. For data quality control, each real-time PCR amplification and dissociation curve should be checked for amplifying one amplicon only, i.e., only one peak should be observed in the melting curve analysis. If the quantification cycle (Cq) for the specific signal exceeded more than 40 cycles or the conducted no reverse transcription control, the miRNA may be considered not detected and the data point may be omitted in the down-stream data analysis.
2. Here, we will briefly mention the comparative quantification "$\Delta\Delta C_T$" method, also known as the $2^{-\Delta\Delta C_T}$ method, which is a more simplistic alternative to the many more sophisticated methods provided in different softwares. The "$\Delta\Delta C_T$" method is applied as follows: the C_T values for all samples (both samples

of interest and reference samples) are extracted and the ΔC_T is calculated as the difference in C_T value between miRNA target and endogenous control [$\Delta C_T = C_{T\text{ (target microRNA)}} - C_{T\text{ (endogenous control)}}$]. The $\Delta\Delta C_T$ is calculated as the $\Delta C_{T\text{ (sample of interest)}} - \Delta C_{T\text{ (reference sample)}}$. Normalization of target gene expression in sample of interest is determined as $2^{-\Delta\Delta C_T}$. The normalized level of expression in the reference sample is set to 1 and the change of target expression is determined as follows: fold change in target miRNA expression = 1 − normalized target miRNA expression in the sample of interest.

3. Subsequent to the first raw data checking, data are ready for normalization. For a comprehensive discussion on this subject, see http://www.gene-quantification.com, IUL Biotechnology Series, A–Z of Quantitative PCR (edited by Stephen A. Bustin), or http://www.exiqon.com/microrna-real-time-pcr.

3.4. Experimental Data Example

1. The experimental setup of the quantitative RT-PCR was run according to the miRCURY LNA™ microRNA PCR System protocol on a Lightcycler 480 system from Roche Applied Science. Total RNA was purified from four different human cell lines (DU-145, MCF-7, PC3, and T-47D). For each total RNA sample, duplicates were made of the first-strand cDNA synthesis, and all quantitative PCR amplifications were made corresponding to four technical replicates. Negative controls, i.e., no RT enzyme and no template, were also included in the PCR.

2. First, a panel of nine potential, stably expressed reference genes was tested (Fig. 1). The data were analyzed using the

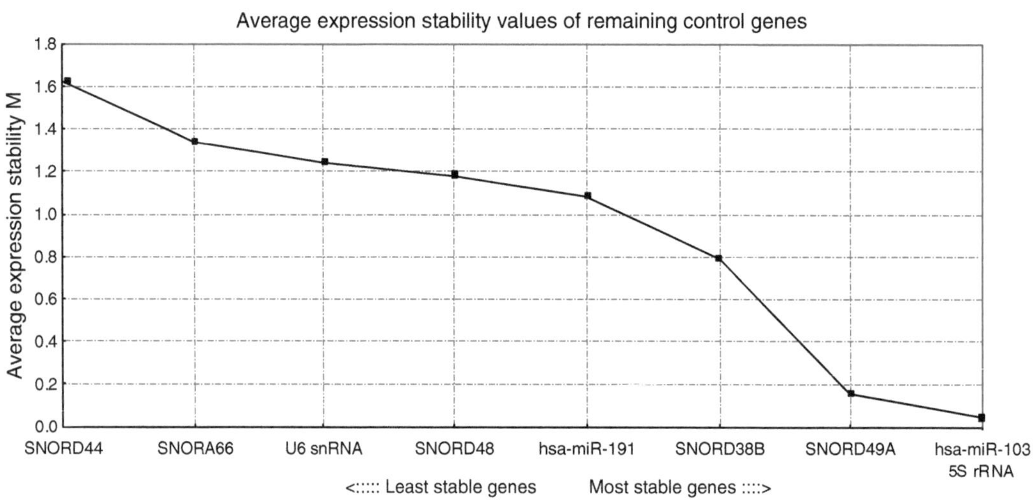

Fig. 1. The graph depicts the average expression stability M of the panel of endogenous controls. The software algorithm GeNorm (13) defines a gene-stability measure M as the average pair-wise variation between a particular gene and all other control genes. In the specific case, the most stably transcribed genes SNORD49A, 5S rRNA, and hsa-miR-103 were selected for normalization.

software GeNorm (13), and three stable reference assays were selected to be included in the miRNA expression profiling experiments.

3. The miRNA primer sets and the three reference genes were run in parallel. The PCR efficiencies were calculated from the sigmoid amplification curves and for each miRNA assay and the mean efficiency were found by averaging the single curve efficiencies measured by the LinRegPCR software (20). The expression profiling analysis was performed using the GenEx software ver. 4.4.2.308 (MultiD Analyses AB). All data were corrected individually for assay efficiency, averaged by technical repeats, and normalized to the three reference genes. The relative values are displayed as log2-transformed mean-centered values (Fig. 2).

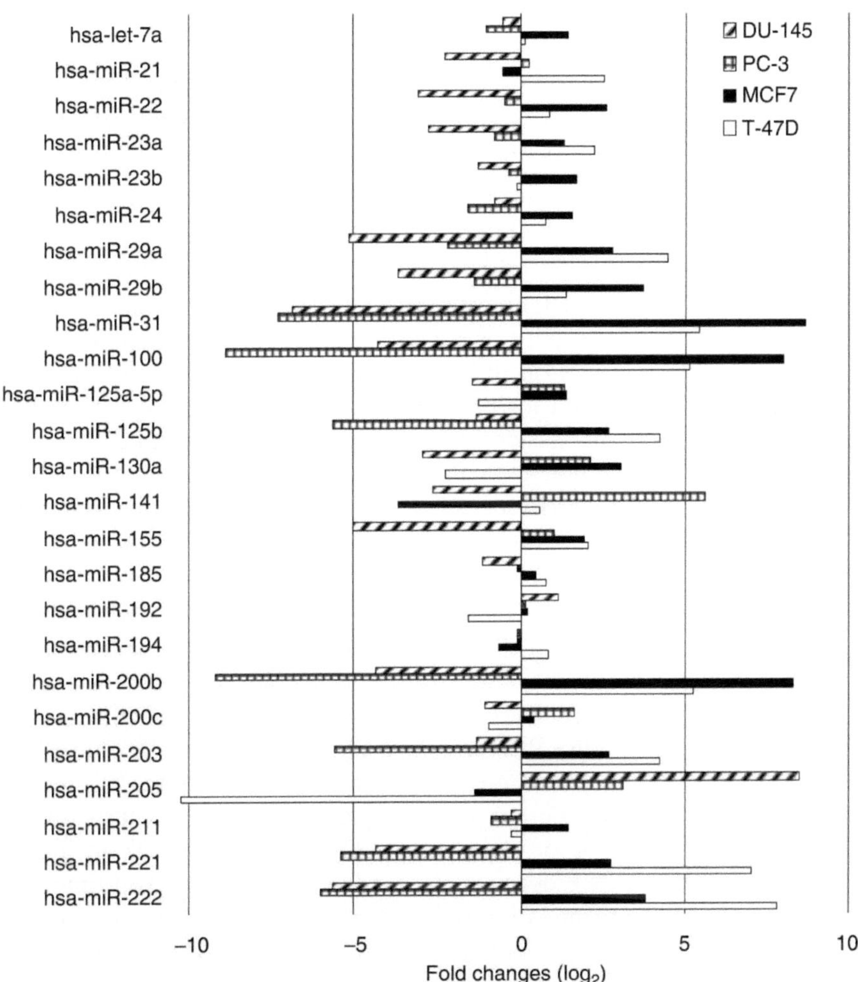

Fig. 2. The expression profiles of different miRNAs in four different cell lines (DU-145, MCF-7, PC3, and T-47D). The cell lines are all members of the NCI60 cancer cell line panel.

4. Notes

1. The isolated RNA from a sample should be a quantitative recovery of small RNAs and should not result in a substantial loss of the small RNA fraction. If necessary, treat the RNA preparations with DNases to remove contaminating genomic DNA that may interfere with the real-time PCR results.
2. Purified total RNA should be dissolved in nuclease-free water at a stock concentration of at least 1 μg/μl in low nucleic acid-binding plasticware. For long-time storage of RNA, it is recommended to precipitate the RNA with ethanol and keep the RNA at –20 or –80°C. Avoid repeated freezing and thawing cycles of the RNA.
3. Keep the reagents and reactions on ice (or at 4°C) as much as possible during the protocol (apart from steps specifically involving raised temperatures). It is recommended that the equipment (e.g., centrifuges and PCR blocks) is cooled before use to avoid exposure to increased temperatures for significant periods of time.
4. To avoid loss of sample, it is recommended that the first-strand cDNA reaction is performed in nuclease-free low nucleic acid-binding (siliconized) micro-centrifuge tubes.
5. In order to minimize the risk of PCR contamination, a working space or a room separated from the space used for RNA isolation and post-analysis of PCR amplicons should be established for first-strand synthesis and the real-time PCR using the master mix setup. Use of pipettes and equipments free of templates is also recommended in order to avoid PCR contaminations. Also, establishment of a daily "one-way" working routine where the template-free setup work is initiated prior to the work with RNA purification, and PCR amplicons can lower the risk of contamination is recommended.
6. Many real-time PCR instruments will produce reliable results only when a passive reference dye such as ROX is added to the PCR. The reference dye is used to normalize signals from individual PCR wells to enable comparison of real-time PCR amplification signals across an entire PCR plate. It is recommended to examine whether your real-time PCR instrument has this type of requirement. The amount of ROX to include in the PCR depends on the requirements of the real-time PCR instrument and must be adjusted accordingly. In general, it is recommended to follow the supplier's instructions for preparation and concentrations of ROX solutions. Typically, real-time PCR instruments that allow excitation at individual wavelengths for individual dyes (most filter wheel based instruments) require

less ROX than instruments that use a single excitation wavelength for all fluorophores (most laser based instruments use excitation at 488 nm).

7. The ability of Uracil-DNA Glycosylase (UDG) to cleave the *N*-glycosidic bond between the uracil base and the phosphodiester backbone of DNA has traditionally been used to avoid PCR carryover. In order to remove DNA from previous PCRs, it is recommended to include UDG in your PCR and replace dTTP with dUTP. The SYBR® Green master mix (vial) is suited for the UDG-based prevention of carryover contamination.

Acknowledgments

We thank Ms. Mette Carlsen Mohr and Ms. Madeline Ek for their skilled technical assistance. We also thank Rolf Søkilde for purification of the total RNA preparations. Finally, we thank Liselotte Kahns for the development of the endogenous control assays.

References

1. Kim, N., Han, J., and Siomi, M.C. (2009) Biogenesis of small RNAs in animals. Nature Reviews *Molecular Cell Biology* **10**, 126–139.
2. Dieffenbach, C.W., Lowe, T.M., and Dveksler, D.S. (1993) General concepts for PCR primer design *Genome Res.* **3**, S30–S37.
3. Raymond, C.K, Roberts, B.S., Garrett-Engele, P., Lim, L.P., and Johnson, J.M. (2005) *RNA* **11**, 1737–1744.
4. Chen C., Ridzon D.A., Broomer, A.J., Zhou, Z., Lee, D.H. et al. (2005) Real-time quantification of microRNAs by stem-loop RT-PCR. *Nucleic Acids Res.* **33**, e179.
5. Wengel,J., Koshkin,A., Singh,S.K., Nielsen,P., Meldgaard,M., et al. (1999) LNA (Locked Nucleic Acid). *Nucleosides & Nucleotides*, **18**, 1365–1370.
6. Bustin, S.A., Benes, V., Nolan, T., Pfaffl, M.W. (2005) Quantitative real-time RT-PCR a perspective. *J. Mol. Endocrinol.* **34**, 597–601.
7. Schmittgen, T.D., Jiang, J., Liu, Q., and Yang, L. (2005) A high-throughput method to monitor the expression of microRNA precursor. *Nucleic Acids Res* **32**, e43.
8. Wittwer, C.T., Herrmann, M.G., Moss, A.A., and Rasmussen R.P. (1997) Continuous fluorescence monitoring of rapid cycle DNA amplification. *Biotechniques* **22**, 130–138.
9. Holland, P.M., Abramson, R.D., Watson, R., and Gelfand, D.H. (1991) Detection of specific polymerase chain reaction product by utilizing the 5'– – 3' exonuclease activity of Thermus aquatius DNA polymerase. *Proc Natl Acad Sci USA* **88**, 7276–7280.
10. Rozen, S. and Skaletsky, H.(2000) Primer3 on the WWW for general users and for biologist programmers. Bioinformatics Methods and Protocols in the series *Methods in Molecular Biology*. Humana Press, Totowa, NJ, **132**, 365–386.
11. Wu H, Neilson JR, Kumar P, Manocha M, Shankar P, et al. (2007) miRNA Profiling of Naïve, Effector and Memory CD8 T Cells. *PLoS ONE* **2**, e1020.
12. Griffiths-Jones, S. (2004) The microRNA Registry. *Nucleic Acids Res.* **32**, Database issue D109–D111.
13. Vandesompele, J., De Preter, K., Pattyn, F., Poppe, B., Van Roy, N., De Paepe, A., Speleman, F. (2002) Accurate normalization of real-time quantitative RT-PCR data by geometric averaging of multiple internal control genes. *Genome Biology* **3**, 34.1–34.11.
14. Andersen C.L., Jensen, J.L. and Ørentoft, T.F., (2004) Normalization of Real-Time Quantitative Reverse Transcription-PCR Data: A Model-Based Variance Estimation Approach to Identify Genes Suited for Normalization, Applied to Bladder and Colon Cancer Data Sets. *Cancer Res.* **64**, 5245–5250.

15. Pfaffl, T.W., Tichopad, A., Prgomet, C., and Neuvians, T.P., (2004) Determination of stable housekeeping genes, differentially regulated target genes and sample integrity: BestKeeper – Excel-based tool using pair-wise correlations. *Biotechnology Letters* **26**, 509–515.
16. Pfaffl, T.W., Horgan, G.W., and Dempfle, L. (2002) *Nucleic Acids Res.* **30**, e36.
17. Chomczynski, P. and Sacchi, N. (1987) Single-step method of RNA isolation by acid guanidinium thiocyanate-phenol-chloroform extraction. *Anal Biochem* **162**,156–159.
18. Peppel, K. and Baglioni, C.A. (1990) Simple and fast method to extract RNA from tissue culture cells. *Biotechniques* **9**, 711–713.
19. Fleige, S. and Pfaffl, M.W. (2006) RNA integrity and the effect on the real-time qRT-PCR performance. *Molecular Aspects of Medicine* **27**, 126–139.
20. Ramakers, C., Ruijter, J.M., Deprez, R.H., and Moorman, A.F. (2003) Assumption-free analysis of quantitative real-time polymerase chain reaction (PCR) data. *Neurosci Lett* **339**, 62–66.

Chapter 5

Detection of Small RNAs and MicroRNAs Using Deep Sequencing Technology

Ericka R. Havecker

Abstract

Small RNAs (sRNAs) have made a large impact on many recent scientific discoveries. MicroRNAs (miRNAs) are a type of sRNA molecule and, although usually just 20–22 nucleotides in length, they are potent regulators of gene expression. Therefore, identification of miRNAs and profiling their abundance are fundamental to understanding an organism's or tissue's gene regulatory network. Next-generation sequencing methods have allowed researchers to quickly sequence and profile sRNA populations. This chapter describes a cloning procedure to identify the sRNAs or miRNAs present in an RNA sample.

Key words: MicroRNAs, Small RNAs, RNA silencing, Next-generation sequencing, Cloning

1. Introduction

RNA silencing is a highly conserved RNA regulatory system, characterized by the presence of small RNA (sRNA) molecules. These guide mRNA cleavage, translational suppression, or epigenetic DNA modifications in a sequence-specific manner (reviewed in ref. 1). sRNAs are divided into subcategories based on their biogenesis and structural features. MicroRNAs (miRNAs) are a specific subcategory of sRNA molecule. They are often highly conserved, 20–22 nucleotides (nt) in length and regulate mRNA transcripts directly (reviewed in ref. 2, 3). However, throughout this protocol the term sRNA is used, as the described procedure can be used to clone most categories of sRNA molecules, including, but not limited to miRNAs.

Most sRNA molecules derive from double-stranded RNA (dsRNA). dsRNA can form through a self-complementary RNA that folds to produce a hairpin structure, as in the case of miRNAs.

Tamas Dalmay (ed.), *MicroRNAs in Development: Methods and Protocols*, Methods in Molecular Biology, vol. 732, DOI 10.1007/978-1-61779-083-6_5, © Springer Science+Business Media, LLC 2011

Alternatively, the dsRNA can be produced from base-pairing of overlapping and complementary RNA transcripts or from a single-stranded RNA acted upon by an RNA-dependent RNA polymerase. During the RNA silencing process, the dsRNA is cleaved into sRNA duplex molecules by an RNAseIII-type enzyme (Dicer or Dicer-like in plants). Cleavage of dsRNA by Dicer/Dicer-like produces sRNAs with a characteristic 5′ monophosphate and 3′ hydroxyl moiety. The described cloning protocol takes advantage of these characteristics. If the sRNA molecule is known to lack the 5′ monophosphate, for example in the case of secondary 5′ triphosphate sRNA molecules, an alternative cloning protocol must be utilized (4–6).

This protocol is based on the sequential ligation of adapters to the 5′ and 3′ ends of a 5′ monophosphate sRNA (7, 8). After the ligations, reverse transcription of the adapter-ligated sRNA molecules takes place and is followed by PCR amplification of the library. The sRNA purification, adapter ligations, and library amplification step are each followed by polyacrylamide gel purification. Once the amplified library has been gel purified, it is ready for high-throughput deep sequencing.

At the time of writing, three different high-throughput deep sequencing platforms are available for miRNA discovery and profiling: Illumina, 454, and ABI solid (reviewed in ref. 9). Although all platforms require adapter ligation to sRNA molecules, each platform uses different adapter sizes and sequences. Adapter sequences are specific for a given technology and cannot be interchanged. Each sequencing platform has different advantages, which chiefly lie in the number of sRNAs that can be profiled, the sequencing error rate of the technology and its cost. The following protocol guides the user toward gel purification steps that are consistent with the size of the Illumina adapter sequences. If another technology is used, the basic cloning procedure is the same, although the user must recalculate the predicted ligation and amplification sizes and adjust the protocol accordingly.

2. Materials

Each of the method's subdivisions generally uses the same reagents, so all reagents and recipes are listed below (see Note 1).

2.1. Polyacrylamide Gel Casting, Electrophoresis, and Visualization

1. Urea.
2. 1× TBE running buffer: diluted from a 5× TBE stock with sterilized, deionized water. Store at room temperature.
3. 40% Acrylamide (19:1 acrylamide:bisacrylamide). Unpolymerized acrylamide and bisacrylamide are potent neurotoxins. Wear appropriate gloves and store solutions at 4°C.

4. 10% Ammonium persulfate prepared freshly in water. Aliquots can be stored at –20°C.
5. *N,N,N,N'*-tetramethyl-ethylenediamine (TEMED).
6. Gel-loading buffer II (Ambion, Inc.).
7. Generuler 100 bp ladder (Fermentas). Mix with standard DNA loading buffer and load 0.5 μg per lane.
8. PCR 20 bp low ladder (SIGMA). Mix with standard DNA loading buffer and load 0.5 μg per lane.
9. Ethidium bromide: this is a toxic mutagen. Wear appropriate gloves when working with solutions that contain ethidium bromide.

2.2. sRNA Purification, Adapter Ligation, RT-PCR, and PCR

1. mirVANA microRNA isolation kit (Ambion, Inc.).
2. Illumina sRNA cloning adapter/oligo kit (Illumina, Inc.).
3. T4 RNA ligase with 10× RNA ligation buffer (Promega).
4. RNaseOUT (Invitrogen).
5. SuperScript II or SuperScript III first-strand cDNA synthesis kit includes 5× first-strand buffer and 100 mM DTT (Invitrogen).
6. dNTPs diluted to 12.5 and 2.5 mM.
7. Phusion DNA polymerase kit includes 5× Phusion high-fidelity buffer (New England Biolabs, Inc.).
8. Nuclease-free water (from Ambion, Inc.).

2.3. Gel Shredding, Elution, and Precipitation of Nucleic Acids

1. 21-Gage needles. Use caution when puncturing microfuge tubes with a needle. Discard the needle when the tip bends.
2. Spin-X Cellulose Acetate Filters (2 mL, 0.45 μm).
3. RNase-free 2 mL microfuge tube (Ambion, Inc.).
4. Nonstick RNase-free 0.5 mL microfuge tube (Ambion, Inc.).
5. 1.5 mL siliconized Eppendorf tubes (SIGMA).
6. Glycoblue (15 mg/mL; Ambion, Inc.).
7. Pellet Paint NF Co-precipitant (Novagen). Store at -20°C, but prior to use thaw at room temperature and vortex for 3–5 min.
8. 100% Isopropanol.
9. 0.3 M sodium chloride. Sterilize by autoclaving, store at room temperature.
10. 3 M sodium acetate pH 5.2. Sterilize by autoclaving, store at room temperature.
11. 100% Ethanol.

2.4. Recipes (see Note 2)

1. 5× TBE: 54 g Tris base, 27.5 g boric acid, 20 mL 0.5 M EDTA, pH 8.0. Adjust volume to 1 L with water and sterilize by autoclaving. Store at room temperature.
2. Gel-loading solution: 10 mL deionized formamide; 200 μL 0.5 M EDTA, pH 8.0, 1 mg xylene cyanol FF, and 1 mg bromophenol blue. Store up to 1 year at 4°C.
3. DNA extraction buffer: 10 mM Tris–glycine, pH 8.0, 10 mM $MgCl_2$, 50 mM NaCl. Sterilize by autoclaving, store at room temperature.
4. Resuspension buffer: 10 mM Tris–glycine, pH 8.5. Sterilize by autoclaving, store at room temperature.
5. 15% TBU gel: 5.25 g urea, 2.5 mL 5× TBE, 4.7 mL 40% acrylamide (19:1 acrylamide:bisacrylamide). Add water to 12.5 mL and dissolve urea by heating solution in a water bath. Add 87.5 μL 10% APS and 4.4 μL TEMED. Pour immediately (see Note 3).
6. 10% TBU gel: 5.25 g urea, 2.5 mL 5× TBE, 3.125 mL 40% acrylamide (19:1 acrylamide:bisacrylamide). Add water to 12.5 mL and dissolve urea by heating solution in a water bath. Add 87.5 μL 10% APS and 4.4 μL TEMED. Pour immediately (see Note 3).
7. 6% TBE gel: 5 mL 5× TBE, 3.75 mL 40% acrylamide (19:1 acrylamide:bisacrylamide). Add water to 25 mL. Add 175 μL 10% APS and 8.8 μL TEMED. Pour immediately (see Note 4).
8. 10% TBE gel: 5 mL 5× TBE, 6.25 mL 40% acrylamide (19:1 acrylamide:bisacrylamide). Add water to 25 mL. Add 175 μL 10% APS and 8.8 μL TEMED. Pour immediately (see Note 4).
9. Oligo nt ladders. To prepare 15–30, 40–60, and 60–80 nt oligo ladder, obtain oligos of the appropriate size, their sequences are unimportant. Combine 250 μL gel-loading buffer, 235 μL water, and 7.5 μL of each 100 μM oligo stock. Denature by heating to 65°C for 10 min, then snap-cool on ice and centrifuge to collect at the bottom of microfuge tube. The oligo ladder dilutions are stored at -20°C.

3. Methods

The sRNA content within a tissue varies, and thus optimization of the amount of total RNA may need to be completed. From tissues with a high proportion of sRNAs, gel purification of the sRNA fraction from 10 μg of total RNA (1 μg/μL) is recommended (see Note 5). In addition, phenol–chloroform-based RNA extraction procedures of total RNA are recommended as they provide the highest yield of sRNAs. Avoid lithium chloride-precipitated RNA samples as sRNAs do not co-precipitate with long RNA transcripts.

3.1. Purify sRNAs from Total RNA

1. Prepare a 1× TBU 15% denaturing polyacrylamide gel (see Note 6).
2. Prerun the gel for 30 min at 200 V. After prerunning but prior to loading, wash the wells with 1× TBE.
3. Mix 10 μL (10 μg) of total RNA with 10 μL of 2× gel-loading buffer II (see Note 7) in an RNase-free microfuge tube.
4. Heat the sample to 65°C for 5 min, then snap-cool on ice. Centrifuge to collect the sample to the bottom of the microfuge tube.
5. Load 10 μL of the denatured 15–30 nt oligo ladder. Leave one lane empty between the sample and the size marker to avoid contamination. If loading multiple samples, load each sample with its own 15–30 nt oligo ladder and separated from the ladder by at least one lane.
6. Run the gel at 150 V until the bromophenol blue dye migrates to the bottom of the gel.
7. Disassemble the gel apparatus. If multiple samples are contained within one gel, using a clean razor blade or scalpel, cut the gel into its appropriate portions (containing the oligo ladder and sample). Stain the entire gel or gel portions by soaking in separate 50 mL aliquots of 1× TBE containing 0.25 μg/mL ethidium bromide for 5 min.
8. Visualize the oligonucleotide size markers on a 360-nm UV transilluminator (see Note 8). Using a clean razor blade or scalpel, excise a gel slice in the sample lane that corresponds to 15–30 nt. This gel slice contains purified sRNA molecules. Transfer it to a 0.5-mL RNase-free microfuge tube whose bottom has been punctured three times by a 21-gage needle.
9. Place the 0.5-mL punctured tube containing the excised gel slice into a 2-mL RNase-free round-bottom microfuge tube. Centrifuge the gel slice through the punctured holes into the 2-mL tube at full speed for 2 min.
10. Remove and discard the empty or nearly empty 0.5 mL tube. Add 500 μL sterile 0.3 M NaCl to the 2-mL tube containing the shredded gel debris.
11. Elute the RNA from the gel debris by rotation of the tube overnight at 4°C.
12. Transfer the eluate and the gel debris into a Spin-X Cellulose Acetate filter. Centrifuge at full speed for 2 min.
13. Wash the gel debris once more by adding 100 μL 0.3 M NaCl to the debris and centrifuge for 2 min. Collect the 100-μL

eluate in the same 2 mL microfuge tube. Remove and discard the Spin-X column containing the gel debris.

14. Precipitate the purified sRNA fraction present within the eluate with an equal volume (approximately 600 μL) of 100% isopropanol and 3 μL glycoblue. Incubate at −80°C for 15–30 min (see Note 9).
15. Pellet the purified sRNAs by centrifugation at 14,000 rpm (18,407 *g*) for 25 min at 4°C.
16. Carefully remove and discard the supernatant. Wash the pellet with 750 μL of room temperature 75% ethanol. Allow the pellet to air-dry and resuspend in 5.7 μL of RNase-free water. The purified sRNAs can be stored at −80°C.

3.2. 5′ Adapter Ligation and Purification (see Note 10)

1. Heat the purified sRNA as described in step 16 of Subheading 3.1 for 30 s at 90°C and then snap-cool on ice. Pulse spin to collect contents at the bottom of a microfuge tube.
2. Set up the 5′ adapter ligation reaction in a 1.5-mL RNase-free siliconized microfuge tube as follows (see Note 11):

Purified miRNA (from Subheading 3.1, step 16)	5.7 μL
5′ RNA adaptor (10 μM)	1.3 μL
10× RNA ligation buffer	1 μL
T4 RNA ligase (10 U/μL)	1 μL
RNaseOUT (40 U/μL)	1 μL

3. Incubate at 37°C for 1 h (see Note 12).
4. Terminate the 5′ adapter ligation reaction by adding 10 μL 2× gel-loading buffer II (see Note 7).
5. Prerun a 15% TBU polyacrylamide gel for 30 min at 200 V (see Note 6). After prerunning but prior to loading, wash the wells with 1× TBE.
6. Heat the ligated sample/loading buffer mixture to 65°C for 5 min, then snap-cool on ice. Centrifuge to collect the sample to the bottom of the microfuge tube.
7. Load the 10 μL of a prepared 40–60 nt oligo ladder followed by an empty well and then the 5′ adapter-ligated sample (see Note 13).
8. Run the gel at 150 V until the xylene cyanol dye migrates to the bottom of the gel (on a 15% TBU gel, the xylene cyanol migrates to approximately the 40-nt size marker).
9. Disassemble the gel apparatus. If multiple samples are contained within one gel, using a clean razor blade or scalpel, cut the gel into its appropriate portions (containing the oligo ladder and sample). Stain the entire gel or gel portions by

soaking in separate 50 mL aliquots of 1× TBE containing 0.25 μg/mL ethidium bromide for 5 min.

10. Visualize the oligo nt size markers on a 360-nm UV transilluminator (see Note 8). Using a clean razor blade or scalpel, excise a gel slice in the sample lane that corresponds to 40–60 nt. This gel slice contains 5′ adapter-ligated sRNAs. Transfer it to a 0.5-mL RNase-free microfuge tube whose bottom has been punctured three times by a 21-gage needle.
11. Place the 0.5-mL punctured tube containing the excised gel slice into a 2-mL RNase-free round-bottom microfuge tube. Centrifuge the gel slice through the punctured holes into the 2-mL tube at full speed for 2 min.
12. Remove and discard the empty or nearly empty 0.5 mL tube. Add 500 μL sterile 0.3 M NaCl to the 2-mL tube containing the shredded gel debris.
13. Elute the 5′ adapter-ligated sRNAs from the gel debris by rotation of the tube overnight at 4°C.
14. Transfer the eluate and the gel debris into a Spin-X Cellulose Acetate filter. Centrifuge at full speed for 2 min.
15. Wash the gel debris once more by adding 100 μL 0.3 M NaCl to the debris and centrifuge for 2 min. Collect the 100-μL eluate in the same 2 mL microfuge tube. Remove and discard the Spin-X column containing the gel debris.
16. Precipitate the 5′ adapter-ligated sRNA fraction present within the eluate with an equal volume (approximately 600 μL) of 100% isopropanol and 3 μL glycoblue. Incubate at −80°C for 15–30 min (see Note 9).
17. Pellet the purified sRNAs by centrifugation at 14,000 rpm (18,407 *g*) for 25 min at 4°C.
18. Carefully remove and discard the supernatant. Wash the pellet with 750 μL of room temperature 75% ethanol. Allow the pellet to air-dry and resuspend in 6.4 μL of RNase-free water. The purified sRNAs can be stored at −80°C.

3.3. 3′ Adapter Ligation and Purification

1. Heat the purified sRNA as mentioned in step 18 of Subheading 3.2 for 30 s at 90°C and then snap-cool on ice. Pulse spin to collect contents at the bottom of a microfuge tube.
2. Set up the 3′ adapter ligation reaction in a 1.5-mL siliconized tube as follows (see Note 11):

Purified 5′ ligation product (Subheading 3.3, step 18)	6.4 μL
3′ RNA adaptor (10 μM)	0.6 μL
10× RNA ligation buffer	1 μL
T4 RNA ligase (10 U/μL)	1 μL
RNaseOUT (40 U/μL)	1 μL

3. Incubate at 37°C for 1 h (see Note 12).
4. Terminate the 3′ adapter ligation reaction by adding 10 μL 2× gel-loading buffer II (see Note 7).
5. Prerun a 10% TBU polyacrylamide gel for 30 min at 200 V (see Note 6). After prerunning but prior to loading, wash the wells with 1× TBE.
6. Heat the ligated sample/loading buffer mixture to 65°C for 5 min, then snap-cool on ice. Centrifuge to collect the sample to the bottom on the microfuge tube.
7. Load the 10 μL of a prepared 60–80 nt oligo ladder followed by an empty well and then the 5′ adapter-ligated sample (see Note 13).
8. Run the gel at 150 V until the xylene cyanol dye migrates to the bottom of the gel (on a 10% TBU gel, the xylene cyanol migrates to approximately the 55 nt size marker).
9. Disassemble the gel apparatus. If multiple samples are contained within one gel, using a clean razor blade or scalpel, cut the gel into its appropriate portions (containing the oligo ladder and sample). Stain the entire gel or gel portions by soaking in separate 50 mL aliquots of 1× TBE containing 0.25 μg/mL ethidium bromide for 5 min.
10. Visualize the oligo nt size markers on a 360-nm UV transilluminator (see Note 8). Using a clean razor blade or scalpel, excise a gel slice in the sample lane that corresponds to 60–80 nt. This gel slice contains your 3′ adapter-ligated sRNAs. Transfer it to a 0.5-mL RNAse-free microfuge tube whose bottom has been punctured three times by a 21-gage needle.
11. Place the 0.5-mL punctured tube containing the excised gel slice into a 2-mL RNase-free round-bottom microfuge tube. Centrifuge the gel slice through the punctured holes into the 2-mL tube at full speed for 2 min.
12. Remove and discard the empty or nearly empty 0.5-mL tube. Add 500 μL sterile 0.3 M NaCl to the 2-mL tube containing the shredded gel debris.
13. Elute the 3′ adapter-ligated sRNAs from the gel debris by rotation of the tube overnight at 4°C.
14. Transfer the eluate and the gel debris into a Spin-X Cellulose Acetate filter. Centrifuge at full speed for 2 min.
15. Wash the gel debris once more by adding 100 μL 0.3 M NaCl to the debris and centrifuge for 2 min. Collect the 100-μL eluate in the same 2-mL microfuge tube. Remove and discard the Spin-X column containing the gel debris.
16. Precipitate the 3′ adapter-ligated sRNA fraction present within the eluate with an equal volume (approximately

600 μL) of 100% isopropanol and 3 μL glycoblue. Incubate at –80°C for 15–30 min (see Note 9).

17. Pellet the purified sRNAs by centrifugation at 14,000 rpm (18,407 *g*) for 25 min at 4°C.
18. Carefully remove and discard the supernatant. Wash the pellet with 750 μL of room temperature 75% ethanol. Allow the pellet to air-dry and resuspend in 4.5 μL of RNase-free water. The purified sRNAs can be stored at –80°C.

3.4. Reverse-Transcription PCR

1. Subject the adapter-ligated sRNAs to reverse transcription using an oligo antisense to the 3′ adapter to prime the reaction. In a 1.5-mL RNase-free microfuge tube combine:

Purified ligated RNA mentioned in Subheading 3.3, step 18	4.5 μL
siRNA RT-primer (100 μM)	0.5 μL

2. Incubate at 65°C for 10 min, pulse spin to cool.
3. Add the following in order:

5× First-strand buffer	2.0 μL
12.5 mM dNTP mix	0.5 μL
100 mM DTT	1.0 μL
RNaseOUT (40 U/μL)	0.5 μL

4. Incubate at 48°C for 3 min and then add 1.0 μL of SuperScript™ II RT (200 U/μL) (see Note 14).
5. Incubate at 42°C for 1 h (see Note 15).

3.5. PCR Amplification and Gel Purification of the sRNA Library (see Note 16)

1. Amplify the cloned sRNA cDNA library using the following setup but aliquot this 80 μL reaction into four thermocycler tubes (see Note 17):

RT reaction as mentioned in step 5 of Subheading 3.4	4 μL
5× High-fidelity buffer	16 μL
PCR Primer 1 (100 μM)	0.5 μL
PCR Primer 2 (100 μM)	0.5 μL
Phusion polymerase (NEB)	0.8 μL
Nuclease-free water	52 μL

Cycling parameters

Step 1: 98°C for 30 s. Step 2: 98°C for 10 s. Step 3: 58°C for 30 s. Step 4: 72°C for 20 s. Step 5: Go to step 2 for 14 additional cycles (or the empirically determined cycle number). Step 6: 72°C for 5 min. Step 7: 4°C forever.

2. Precipitate the PCR amplifications by combining the aliquots from the same libraries into one 1.5 mL microfuge tube (see Note 18). Add 0.1× (v/v) 3 M sodium acetate pH 5.2, 2.5× (v/v) 100% ethanol and 3 μL glycoblue. For the above reaction, add to 80 μL of the combined PCR, 8 μL 3 M sodium acetate pH 5.2, 200 μL 100% ethanol, and 3 μL glycoblue.
3. Precipitate the amplified library at −80°C for 15–30 min (see Note 9).
4. Pellet the reaction by centrifugation at 14,000 rpm (18,407 *g*) for 25 min.
5. Wash the pellet with 500 μL 70% ethanol.
6. Carefully remove and discard the supernatant. Allow the pellet to air-dry.
7. Resuspend the pellet (containing the PCR amplified library) in 20 μL water. Add 3 μL 10× DNA loading dye containing xylene cyanol and bromophenol blue.
8. Load a 6% TBE polyacrylamide gel with two size markers, a PCR 20-bp low ladder (0.5 μg SIGMA) and a 100-bp ladder (0.5 μg Fermentas 100 bp Generuler ladder). Leave an empty well and load one PCR amplification reaction on two consecutive lanes by splitting the sample into half (see Note 19). Prerunning this gel is unnecessary.
9. Run the gel at 100 V until the xylene cyanol migrates to approximately two-thirds the length of the gel. Xylene cyanol migrates to 106 nt on a 6% polyacrylamide gel.
10. Disassemble the gel apparatus. If multiple samples are contained within one gel, using a clean razor blade or scalpel, cut the gel into its appropriate portions (containing the two size markers and sample). Stain the entire gel or gel portions by soaking in separate 50 mL aliquots of 1× TBE containing 0.25 μg/mL ethidium bromide for 5 min.
11. Visualize the size markers and amplified sRNA library on a 360-nm UV transilluminator (see Note 8). The smallest size marker on the 100-bp ladder is 100 bp. It corresponds to 100 bp on the 20-bp low ladder.
12. Using a clean razor blade or scalpel, excise a gel slice in the sample lane that corresponds to 92 bp (see Note 20). This gel slice contains your amplified sRNA library. Place the gel slice into a 0.5-mL microfuge tube whose bottom has been punctured three times by a 21-gage needle.
13. Place the 0.5-mL punctured tube containing the gel slice into a 2-mL round-bottom microfuge tube and shred the gel slice through the holes and into the 2-mL tube by centrifuging at top speed for 2 min. Remove and discard the punctured 0.5-mL microfuge tube.

14. To elute the amplified sRNA library from the shredded gel slice, add 300 μL extraction buffer to the gel debris in the 2-mL microfuge tube. Shake at 30°C for 1 h and then rotate at 4°C overnight (see Note 21).
15. Transfer the eluate and the gel debris into a Spin-X Cellulose Acetate filter. Centrifuge for 2 min at top speed. Wash the gel debris by adding 100 μL of DNA extraction buffer to the gel debris and then centrifuge 2 min at top speed. Collect the second wash in the same 2-mL microfuge tube.
16. Precipitate the amplified sRNA library with 40 μL 3 M sodium acetate pH 5.2, 1 mL 100% ethanol, and 1 μL Pellet Paint (see Note 22).
17. Pellet the amplified sRNA library by centrifugation at full speed for 25 min.
18. Wash the pellet with 1 mL 70% EtOH and allow to air-dry. Resuspend the pellet in 15 μL resuspension buffer.

3.6. Quantification of the Amplified sRNA Library

The concentration of the amplified sRNA library is critical for accurate next-generation sequencing, and its requirement varies based upon the platform utilized. In addition, as technologies improve, lower concentrations may become sufficient. We recommend obtaining the sRNA library requirements from the center where sequencing will take place. This center may also have a method for obtaining the library concentration. An accurate device for measuring concentrations would include a BioAnalyzer, while traditional spectrophotometer readings are fairly inaccurate. It is also advisable to subject 1–2 μL of the purified sRNA library to electrophoresis on a 10% TBE polyacrylamide gel to visualize and ensure that the 92-bp PCR product has been recovered.

4. Notes

1. We have attempted to list the suppliers we generally use for some enzymes and reagents. However, numerous competitive reagents are often available from other commercial sources and can be substituted.
2. Use deionized, distilled, and autoclaved water in all recipes and protocol steps.
3. 12.5 mL will be sufficient for casting two 0.75 mm gels (we use the BioRad MiniProtean II gel apparatus, but others can substitute). Scale up or down according to the number of gels needed. Wear appropriate gloves when mixing the polyacrylamide gels.

4. 25 mL will be sufficient for casting two 1.5 mm gels (we use the BioRad MiniProtean II gel apparatus, but others can substitute). Scale up or down according to the number of gels needed. Wear appropriate gloves when mixing the polyacrylamide gels.
5. mirVANA sRNA isolation from a total RNA sample can replace the entire sRNA gel purification procedure (Subheading 3.1) by following the manufacturer's protocol. However, after sRNA isolation, the sRNA fraction must be concentrated for compatibility with the 5′ adapter ligation reaction. Concentrate the mirVANA isolated sRNA fraction with 0.1× (v/v) 3 M sodium acetate pH 5.2, 2.5× (v/v) 100% ethanol, and 1 μL glycoblue. Resuspend the sRNA fraction in 5.7 μL per 10 μg total RNA input. For example, if starting with 100 μg total RNA, resuspend the sRNA fraction in 57 μL and use 5.7 μL for Subheading 3.2, step 2.
6. Commercially available gels can be utilized. 10 well, 0.75 mm or 1 mm width is advisable.
7. If 2× gel-loading buffer II is not available, substitute with 10 μL of gel-loading solution (see recipe).
8. Visualization usually requires a 360-nm UV transilluminator. The gel can be protected from RNAse contamination by placing plastic wrap between the gel and the transilluminator. In addition, a blue light transilluminator can replace the UV transilluminator, but visualization may not be as sensitive as with the UV transilluminator.
9. Precipitation can also take place overnight at –20°C.
10. The sizes for gel excision are based on the Illumina adapters. If alternative next-generation sequencing technologies are utilized, the user must calculate the predicted size after ligation (by adding the adapter length to the expected sRNA size) and modify the gel excision parameters and potentially the oligo ladder accordingly.
11. This step can take place in a thermocycler, and one could set up the ligation in a 0.2-mL thermocycler tube.
12. Ligation can take place at +20°C for 6 h followed by incubation at 4°C overnight.
13. If cloning multiple samples at once, it is advisable to run each sample with its own ladder and cut the gel before staining or run each sample on separate gels to avoid contamination amongst samples.
14. SuperScript III can be substituted for SuperScript II.
15. Incubate at 50°C if using SuperScript III.
16. It is advisable to set up a single 20 μL PCR using 15 cycles to empirically determine if 15 cycles are adequate for amplification

of the cloned sRNA library. To do this, set up the PCR as in Subheading 3.5, step 1 but modify the master mix amounts to one-fourth of the numbers given. Subject the PCR to electrophoresis on a 6% 1× TBE gel to determine if the 92-bp product is evident (as in Subheading 3.5, step 8; concentration of PCR product is unnecessary). If the 92-bp product exists, proceed with the amplification of the library as mentioned in step 1 of Subheading 3.5. If the 92-bp product is not visible, try increasing to 20 or 25 cycles to obtain a product. If a strong 92-bp product exists at 15 cycles, the researcher may want to consider reducing the number of cycles so as to not over amplify abundant sRNA species.

17. This protocol uses 4 μL of the RT reaction for the first PCR amplification. If failure occurs, the researcher could try at least once more, but empirical determination of thermocycler number may be necessary.
18. To clone a portion of the sRNA library into a T-overhang vector for manual sequencing, add 1 μL of Taq polymerase to the PCR (Subheading 3.5, step 1) after it has finished. Incubate the reaction at 72°C for 15 min, to add adenosine overhangs to the amplified products. Phusion polymerase does not adenosine overhangs.
19. To avoid contamination, it is advisable to use separate polyacrylamide gels per library.
20. Often, a by-product of this cloning procedure (5′ and 3′ adapters ligated to each other) is visible. For Illumina adapters, this band migrates to approximately 70 nt. Other sizes can sometimes be visualized after amplification, but they are often artifacts and the researcher is advised to excise only the 92-bp band.
21. Elution can occur with incubation at 4°C overnight or at 37°C for 2 h with shaking. 1× NEB Buffer 2 can substitute for the described elution buffer.
22. 3 μL glycoblue can be substituted for Pellet Paint NF Co-precipitant, which is a carrier compatible with fluorescent sequencing.

References

1. Broderson, P. and Voinnet, O. (2006) The diversity of RNA silencing pathways in plants. *Trends Genet.* **22**, 268–280.
2. He, L. and Hannon, G. J. (2004) MicroRNAs: small RNAs with a big role in gene regulation. *Nature Rev. Genet.* **5**, 631.
3. Narry Kim, V. (2005) MicroRNA biogenesis: coordinated cropping and dicing. *Nature Rev. Mol. Cell. Biol.* **6**, 376–385.
4. Baulcombe, D. C. (2007) Amplified silencing. *Science* **315**, 199–200.
5. Pak, J. and Fire, A. (2007) Distinct populations of primary and secondary effectors during RNAi in C. elegans. *Science* **315**, 241–244.
6. Sijen, T., Steiner, F. A., Thijssen, K. L., and Plasterk, R. H. (2007) Secondary siRNAs result from unprimed RNA synthesis and form a distinct class. *Science* **315**, 244–247.

7. Pfeffer, S., Lagos-Quintana, M., and Tuschl, T. (2003) Cloning of small RNA molecules, in *Current Protocols in Molecular Biology* (Ausubel, F. M., Brent, R., Kingston, R. E., Moore, D. D., Seidman, J. G., Smith, J. A., and Struhl, K., eds.) John Wiley & Sons, Inc., Hoboken, NJ, 26.4.1–26.4.18.
8. Chappell, L., Baulcombe, D. C., and Molnar, A. (2005) Isolation and cloning of small RNAs from virus-infected plants, in *Current Protocols in Microbiology* (Coico, R., Kowalik, T., Quarles, J. M., Stevenson, B., and Taylor, R. K., eds.) John Wiley & Sons, Inc., Hoboken, NJ, 16H.12.11–16H.12.17.
9. Mardis, E. R. (2008) Next-generation DNA sequencing methods. *Annu. Rev. Genom. Human Genet.* 9, 387–402.

Chapter 6

Detection of MicroRNAs in Prostate Cancer Cells by MicroRNA Array

Xiaoqing Tang, Xiaohu Tang, Jozsef Gal, Natasha Kyprianou, Haining Zhu, and Guiliang Tang

Abstract

MicroRNAs (miRNAs) are a novel class of small noncoding RNAs that regulate gene expression at the posttranscriptional level and play a critical role in many important biological processes and pathological development. In the past few years, miRNAs have been implicated to play an important role in cancer initiation and development. In this chapter, we describe a protocol for the analysis and characterization of miRNAs in prostate cancer cells using a simple but effective array platform. The array is composed of 553 nonredundant miRNAs encompassing the entire set of known miRNAs in humans and mice. As an example, profiling of miRNAs in four prostate cancer cell lines has revealed that a set of miRNAs were differentially expressed between androgen-dependent and androgen-independent metastatic prostate cancer cells. Among them, the differential expression of miR-205 and miR-200c were further validated by Northern blot analysis in these two types of prostate cancer cells. This comprehensive and easy-to-follow protocol will be useful for studying miRNAs in various cancers and can be readily adapted for miRNA analysis in a variety of human diseases.

Key words: MicroRNA, MicroRNA array, Prostate cancer

1. Introduction

MicroRNAs (miRNAs) are a recently discovered class of small noncoding RNAs that regulate gene expression at the posttranscriptional level in plants and animals (1–3). To date, about 10,867 miRNAs have been identified from 115 organisms, including animals, plants, and additional 19 animal viruses, which are currently deposited in the miRBase registry (http://www.mirbase.org/) (4, 5). From these, 721, 579, and 325 miRNAs have been identified in humans, mice, and rats, respectively.

Tamas Dalmay (ed.), *MicroRNAs in Development: Methods and Protocols*, Methods in Molecular Biology, vol. 732, DOI 10.1007/978-1-61779-083-6_6, © Springer Science+Business Media, LLC 2011

Bioinformatics analysis predicts that miRNAs comprise ~5% of the transcriptome (6) and may regulate the translation of more than one third of human mRNA species (7, 8). An additional independent analysis predicted that 20% of human genes can potentially be regulated by miRNAs (9). These miRNA molecules are highly conserved among organisms and play critical roles in many important biological processes, including cancer development (10–14).

miRNAs are particularly important for understanding the tumorigenesis, progression and metastasis of human cancers (15–20). The expression profiles of miRNAs can be used for the classification, diagnosis, and prognosis of human malignancies (21–23). Cancer-specific miRNA fingerprints have been identified in many types of cancers, including B-CLL (10), breast carcinoma (24), primary glioblastoma (25), hepatocellular carcinoma (26), papillary thyroid carcinoma (27), lung cancer (13, 28, 29), gastric carcinoma, colon carcinoma (30), and endocrine pancreatic tumors (13). Thus, alterations in miRNAs play a critical role in the pathophysiology of many, if not all, human cancers. Furthermore, a miRNA network has been established by analyzing the expression patterns of miRNAs and their targeted pathways from various cancer cells (31).

Prostate cancer is a major contributor to cancer-related mortality in American men of all ages, causing the death of approximately 30,000 men each year (32). Prostate cancer mortality results from metastasis to the bone and lymph nodes and progression from an androgen-dependent to an androgen-independent disease. Although localized prostate cancer is curable, no effective treatment is available for metastatic prostate cancer. Thus, an early detection of the local tumor is important for its treatment and requires further understanding of the mechanisms underlying the emergence of advanced androgen-independent prostate cancer from androgen-dependent tumors. Recent studies have found that miRNAs are involved in prostate cancer progression (13, 33–36). These studies used sensitive miRNA array technologies to determine miRNA expression patterns in androgen-dependent and androgen-independent metastatic prostate cancer cell lines as well as patient samples. Some specific and unique miRNAs have been demonstrated to contribute to the prostate cancer initiation, progression, and metastasis (22, 37–46). Here, we provide a comprehensive protocol for the analysis of miRNA expression in various prostate cancer cell lines using our recently developed miRNA array platform (47). A cluster of miRNAs have been identified to be uniquely associated with different types of prostate cancer cell lines, which provides a basis for further study of the roles of miRNAs in the development of prostate cancer. This protocol can be extended to the analysis of all kinds of other cancer cells and other types of cells/tissues from various human diseases.

2. Materials

2.1. miRNA Array Platform Design

1. miRNA Database (miRBase: http://www.mirbase.org/).
2. 553 selected DNA oligonucleotide probes and the five external controls (MAC1, MAC2, MAC3, MAC4, and MAC5, where MAC represents miRNA array control) were synthesized by Integrated DNA Technologies (IDT) (47).
3. Hybond N^+ Nylon membrane (Amersham).
4. A high-throughput array probe printing robot, Genetix Qpix2.
5. UV crosslinker (Ultra-Violet Products).
6. 20× SSPE: 3 M NaCl, 0.25 M NaH_2PO_4, and 0.02 M EDTA, adjust to pH 7.4 with 10 N NaOH.

2.2. Prostate Cancer Cell Lines and Cell Culture

1. Four established prostate cell lines, BPH-1, LNCaP, DU-145, and PC-3, were chosen for this study. They represent immortalized benign prostate cells (BPH-1), mildly tumorigenic, mildly metastatic, and androgen-dependent malignant cells (LNCaP), and highly tumorigenic, highly metastatic, and androgen-independent malignant cells (PC-3 and DU-145).
2. RPMI-1640 medium (Sigma) supplemented with 10% fetal bovine serum (FBS), 100 μg/ml penicillin, and 100 μg/ml streptomycin.

2.3. Total RNA Isolation and Small RNA Enrichment

1. Ice-cold PBS.
2. Cell scraper.
3. TRIzol (Invitrogen).
4. Chloroform.
5. Isopropanol.
6. 100 and 80% cold ethanol.
7. RNase-free water.
8. Nanodrop1000™ spectrophotometer (Nanodrop Technologies).
9. Formamide loading dye (2×): 98% w/v deionized formamide, 10 mM EDTA, pH 9.0, 0.025% w/v xylene cyanol, and 0.025% w/v bromophenol blue.
10. SequaGel System (National Diagnostics): concentrate, diluent, buffer.
11. TEMED (*N*,*N*,*N*′,*N*′-Tetramethylethylenediamine).
12. 10% APS (Ammonium Persulfate).
13. TBE buffer (20×).

14. Vertical Gel Electrophoresis Apparatus (GIBCO-BRL Model V16).
15. Gel elution buffer: 100 mM TRIS-glycine, pH 7.5, 300 mM NaCl, and 10 mM EDTA.
16. Glycogen (20 mg/ml, Roche).

2.4. Small RNA Dephosphorylation and Radiolabeling

1. Antarctic Phosphatase (AP) (New England Biolabs).
2. T4 polynucleotide kinase (PNK, 10,000 U/ml, New England Biolabs).
3. RNasin (40 U/μl, Promega).
4. Radiochemicals: γ-^{32}P adenosine triphosphate (γ-^{32}P-ATP) (6,000 Ci/mmol, ≥10 mCi/ml, NEN or ICN).
5. Sephadex G-25 chromatography column (Roche).

2.5. Array Membrane Prehybridization and Hybridization

1. 20× SSPE: 3 M NaCl, 0.25 M NaH_2PO_4, and 0.02 M EDTA, adjust to pH 7.4 with 10 N NaOH.
2. 20× SSC: 3 M NaCl and 0.3 M sodium citrate, adjust to pH 7.0 with 1 M HCl.
3. 10% SDS.
4. 50× Denhardt's reagent: 1% w/v Ficoll (type 400, Pharmacia), 1% w/v poly-vinylpyrrolidone, and 1% w/v bovine serum albumin (Fraction V, Sigma).
5. 10 mg/ml Herring sperm DNA (Invitrogen).
6. Formamide.
7. Array membrane made of Hybond N^+ Nylon membrane and spotted with 553 miRNA probes.
8. Hybridization bottles.
9. VWR Hybridization oven.

2.6. Array Image Output and Analysis

1. PhosphorImager screen.
2. PhosphorImager scanner (Typhoon 9400, Molecular Dynamics).
3. Software: ImageQuant TL (GE Healthcare), Cluster 3.0 and Java TreeView.

2.7. Northern Blot and Probe Stripping

1. Trans-Blot Semi-Dry Electrophoretic Transfer Cell apparatus (Bio-Rad).
2. Hybond N^+ Nylon membrane (Amersham).
3. 3 mm Whatman paper (Bio-Rad).
4. UV crosslinker (Ultra-Violet Products).
5. Stripping buffer: 0.1× SSC and 0.5% SDS.

3. Methods

A successful miRNA array depends on the design of an array platform (see Note 1). High-quality RNA preparation is a key step to begin miRNA array analysis (see Note 2).

3.1. The Overall Design of a Simple miRNA Array Platform

1. Analyze the sequences of all human and mouse miRNAs in the miRNA database (miRBase: http://www.mirbase.org/) using an Excel spreadsheet.
2. Select conserved miRNAs that have a stretch of more than 18 continuous identical nucleotides and design one probe for each of such highly conserved miRNA clusters; select those unique miRNAs that have a stretch of less than 18 continuous identical nucleotides and design specific probes for each of them. Altogether 553 probes were selected to detect the entire set of human and mouse miRNAs that was available in the miRBase in August 2006 (see Note 3).
3. Design and synthesize 553 miRNA antisense DNA oligos.
4. Design and synthesize five unique external controls.
5. Design the array map (Table 1) in a 32 row × 18 column format so that the 553 miRNA probes and the five external control small RNA probes can be printed on the array membrane. All the miRNA probes and the control probes are printed in duplicates.
6. Spot 0.25 pmol of each probe in duplicate onto a Hybond N^+ membrane using a printing robot such as the Genetix Qpix2.
7. UV crosslink the spotted membrane twice with 1,200 × 100 μJ/cm^2 for half a minute each time, before and after wetting the membrane with 2× SSPE buffer.

3.2. Total RNA Preparation from Different Prostate Cancer Cells

The prostate cell lines used in this study represent immortalized benign prostate cells (BPH-1), mildly tumorigenic, mildly metastatic, and androgen-dependent malignant cells (LNCaP), and highly tumorigenic, highly metastatic, and androgen-independent malignant cells (PC-3 and DU-145). The differences among these cells will likely provide valuable insights into the role of the differentially expressed miRNAs in prostate cancer tumorigenesis and progression. In particular, most prostate cancers are initially dependent on androgens and regress with antiandrogen therapy. But recurrences after antiandrogen therapy are often androgen-independent and highly metastatic. Studying the androgen-sensitive cancer cell line (LNCaP) and two androgen-refractory cell lines (PC-3 and DU-145) may identify miRNAs that are potentially involved in the response of a prostate cancer cell to androgens.

Table 1
Human and mouse miRNA array layout

#	1	2	3	4	5	6
1	H-miR-1	H-miR-7	H-let-7b	H-let-7c	H-let-7d	M-let-7d*
33	H-miR-26b	H-miR-27b	H-miR-27a	H-miR-28	H-miR-29a	H-miR-29b
65	H-miR-105	H-miR-106a	M-miR-106a	H-miR-106b	H-miR-107	H-miR-122a
97	H-miR-143	H-miR-144	H-miR-145	H-miR-146a	H-miR-146b	H-miR-147
129	H-miR-191	H-miR-191*	H-miR-192	H-miR-193a	H-miR-193b	H-miR-194
161	H-miR-214	H-miR-215	M-miR-215	H-miR-216	H-miR-217	H-miR-218
193	H-miR-302b	H-miR-302b*	M-miR-302b*	H-miR-302c	H-miR-302c*	H-miR-302d
225	H-miR-363	H-miR-363*	H-miR-365	H-miR-367	H-miR-368	H-miR-369-5p
257	H-miR-411	M-miR-411	H-miR-412	H-miR-421	H-miR-422a	H-miR-422b
289	M-miR-465	M-miR-466	M-miR-467a	M-miR-467b	M-miR-467*	M-miR-468
321	H-miR-499	H-miR-500	M-miR-500	H-miR-501	M-miR-501*	H-miR-502
353	H-miR-519a	H-miR-519c	H-miR-519d	H-miR-519e	H-miR-519e*	H-miR-520a
385	M-miR-546	M-miR-547	H-miR-548a	H-miR-548b	H-miR-548c	H-miR-548d
417	H-miR-574	H-miR-575	H-miR-576	H-miR-577	H-miR-578	H-miR-579
449	H-miR-606	H-miR-607	H-miR-608	H-miR-609	H-miR-610	H-miR-611
481	H-miR-638	H-miR-639	H-miR-640	H-miR-641	H-miR-642	H-miR-643
513	M-miR-680	M-miR-681	M-miR-682	M-miR-683	M-miR-684	M-miR-685
545	M-miR-712	M-miR-713	M-miR-714	M-miR-715	M-miR-717	M-miR-718

#	13	14	15	16	17	18
1	H-miR-10a	H-miR-10b	H-miR-15a	H-miR-15b	H-miR-16	H-miR-17-5p
33	H-miR-30e-3p	H-miR-31	H-miR-32	H-miR-33	H-miR-33b	H-miR-34a
65	H-miR-128a	H-miR-129	M-miR-129-3p	H-miR-130a	H-miR-130b	H-miR-132
97	H-miR-152	H-miR-153	H-miR-154	H-miR-154*	H-miR-155	H-miR-181a
129	H-miR-199a*	H-miR-199b	M-miR-199b	H-miR-200a	H-miR-200a*	H-miR-200b
161	M-miR-290	M-miR-291a-5p	M-miR-291b-3p	M-miR-291a-3p	M-miR-292-5p	M-miR-292-3p
193	H-miR-326	H-miR-328	H-miR-329	H-miR-330	H-miR-331	H-miR-335
225	H-miR-374	H-miR-375	H-miR-376a	H-miR-376a*	M-miR-376b	M-miR-376b*
257	H-miR-431	H-miR-432	H-miR-432*	H-miR-433	M-miR-433-5p	M-miR-434-3p
289	H-miR-485-5p	H-miR-485-3p	H-miR-486	H-miR-487a	H-miR-487b	M-miR-488
321	H-miR-509	H-miR-510	H-miR-511	H-miR-512-5p	H-miR-512-3p	H-miR-513
353	H-miR-521	H-miR-522	H-miR-523	H-miR-524	H-miR-524*	H-miR-525
385	H-miR-554	H-miR-555	H-miR-556	H-miR-557	H-miR-558	H-miR-559
417	H-miR-586	H-miR-587	H-miR-588	H-miR-589	H-miR-590	H-miR-591
449	H-miR-618	H-miR-619	H-miR-620	H-miR-621	H-miR-622	H-miR-623
481	H-miR-650	H-miR-651	H-miR-652	H-miR-653	H-miR-654	H-miR-655
513	M-miR-692	M-miR-693	M-miR-694	M-miR-695	M-miR-696	M-miR-697
545						

7	8	9	10	11	12	#
H-let-7e	H-let-7f	H-let-7g	H-let-7i	H-miR-9	H-miR-9*	32
H-miR-29c	H-miR-30a-5p	H-miR-30a-3p	H-miR-30c	H-miR-30d	H-miR-30e-5p	64
H-miR-124a	H-miR-125a	H-miR-125b	H-miR-126	H-miR-126*	H-miR-127	96
H-miR-148a	H-miR-148b	H-miR-149	H-miR-150	H-miR-151	M-miR-151	128
H-miR-195	H-miR-196a	H-miR-196b	H-miR-197	H-miR-198	H-miR-199a	160
H-miR-219	H-miR-220	H-miR-221	H-miR-222	H-miR-223	H-miR-224	192
H-miR-320	M-miR-322	H-miR-323	H-miR-324-5p	H-miR-324-3p	H-miR-325	224
H-miR-369-3p	H-miR-370	H-miR-371	H-miR-372	H-miR-373	H-miR-373*	256
H-miR-423	H-miR-424	H-miR-425-5p	H-miR-425-3p	H-miR-429	M-miR-429	288
M-miR-469	M-miR-470	M-miR-471	H-miR-483	M-miR-483	H-miR-484	320
H-miR-503	H-miR-504	H-miR-505	H-miR-506	H-miR-507	H-miR-508	352
H-miR-520a*	H-miR-520c	H-miR-520d	H-miR-520d*	H-miR-520e	H-miR-520g	384
H-miR-549	H-miR-550	H-miR-551a	H-miR-551b	H-miR-552	H-miR-553	416
H-miR-580	H-miR-581	H-miR-582	H-miR-583	H-miR-584	H-miR-585	448
H-miR-612	H-miR-613	H-miR-614	H-miR-615	H-miR-616	H-miR-617	480
H-miR-644	H-miR-645	H-miR-646	H-miR-647	H-miR-648	H-miR-649	512
M-miR-686	M-miR-687	M-miR-688	M-miR-689	M-miR-690	M-miR-691	544
M-miR-719	M-miR-720	M-miR-721				576

19	20	21	22	23	24	#
H-miR-17-3p	M-miR-17-3p	H-miR-18a	H-miR-18a*	H-miR-19a	H-miR-19b	32
H-miR-34b	H-miR-34c	H-miR-92	H-miR-92b	H-miR-93	H-miR-95	64
H-miR-133a	M-miR-133a*	H-miR-134	H-miR-135a	H-miR-135b	H-miR-136	96
H-miR-181a*	H-miR-181b	H-miR-181c	H-miR-181d	H-miR-182	H-miR-182*	128
H-miR-200c	M-miR-201	H-miR-202	H-miR-202*	H-miR-203	H-miR-204	160
M-miR-293	M-miR-294	M-miR-295	H-miR-296	M-miR-297	M-miR-297b	192
H-miR-337	H-miR-338	H-miR-339	H-miR-340	M-miR-341	H-miR-342	224
M-miR-376c	H-miR-377	H-miR-378	H-miR-379	H-miR-380-5p	H-miR-380-3p	256
M-miR-434-5p	H-miR-448	H-miR-449	M-miR-449b	H-miR-450	M-miR-450b	288
M-miR-488	M-miR-488*	H-miR-489	M-miR-489	H-miR-490	H-miR-491	320
H-miR-514	H-miR-515-5p	H-miR-515-3p	H-miR-516-5p	H-miR-516-3p	H-miR-517a	352
H-miR-525*	H-miR-526b	H-miR-526b*	H-miR-526c	H-miR-527	H-miR-532	384
H-miR-560	H-miR-561	H-miR-562	H-miR-563	H-miR-564	H-miR-565	416
H-miR-592	H-miR-593	H-miR-594	H-miR-595	H-miR-596	H-miR-597	448
H-miR-624	H-miR-625	H-miR-626	H-miR-627	H-miR-628	H-miR-629	480
H-miR-656	H-miR-657	H-miR-658	H-miR-659	H-miR-660	H-miR-661	512
M-miR-698	M-miR-699	M-miR-700	M-miR-701	M-miR-702	M-miR-703	544
				MAC1		576

(continued)

Table 1 (continued)

#	25	26	27	28	29	30
1	H-miR-20a	H-miR-20b	H-miR-21	H-miR-22	H-miR-23b	H-miR-24
33	H-miR-96	H-miR-98	H-miR-99a	H-miR-99b	H-miR-100	H-miR-101
65	H-miR-137	H-miR-138	H-miR-139	H-miR-140	M-miR-140*	H-miR-141
97	H-miR-183	H-miR-184	H-miR-185	H-miR-186	H-miR-187	H-miR-188
129	H-miR-205	H-miR-206	M-miR-207	H-miR-208	H-miR-210	H-miR-211
161	M-miR-298	M-miR-299	M-miR-300	H-miR-301	H-miR-299-5p	H-miR-299-3p
193	M-miR-344	H-miR-345	H-miR-346	M-miR-346	M-miR-350	M-miR-351
225	H-miR-381	H-miR-382	H-miR-383	M-miR-383	H-miR-384	H-miR-409-5p
257	M-miR-450b*	H-miR-451	H-miR-452	H-miR-452*	H-miR-453	H-miR-455
289	H-miR-492	H-miR-493-5p	H-miR-493-3p	H-miR-494	H-miR-495	H-miR-496
321	H-miR-517c	H-miR-517*	H-miR-518a	H-miR-518b	H-miR-518c*	H-miR-518d
353	H-miR-539	M-miR-540	M-miR-541	H-miR-542-5p	H-miR-542-3p	M-miR-543
385	H-miR-566	H-miR-567	H-miR-568	H-miR-569	H-miR-570	H-miR-571
417	H-miR-598	H-miR-599	H-miR-600	H-miR-601	H-miR-602	H-miR-603
449	H-miR-630	H-miR-631	H-miR-632	H-miR-633	H-miR-634	H-miR-635
481	H-miR-662	H-miR-663	M-miR-669a	M-miR-669b	M-miR-669c	M-miR-677
513	M-miR-704	M-miR-705	M-miR-706	M-miR-707	M-miR-708	M-miR-709
545	MAC2		MAC3		MAC4	

Complementary DNA oligos (0.25 pmol) were spotted in the corresponding locations for each of the selected miRNAs. Five external array control probes (0.25 pmol; MAC1, 2, 3, 4, and 5) were spotted at the bottom right of the membranes (*miR* MicroRNA, *H-miR* Human miRNA, and *M-miR* Mouse miRNA).

Identification of such miRNAs will be valuable for the mechanistic understanding of the androgen responsiveness of prostate cancer.

1. Rinse cells with ice-cold PBS once. Lyse cells directly in a culture dish by adding 1 ml of TRIzol reagent per 6 cm diameter dish and scraping with cell scraper. Transfer the lysates to tubes. Pass the cell lysates through a syringe with a needle several times. Incubate the homogenized sample for 5 min at room temperature to facilitate the complete dissociation of nucleoprotein complexes.
2. Add 0.2 ml chloroform per 1 ml TRIzol. Cap sample tubes securely. Vortex samples vigorously for 15 s and incubate them at room temperature for 5 min. Centrifuge the samples at 12,000 × *g* for 15 min at 4°C. Following centrifugation, the mixture separates into lower red, phenol–chloroform phase, an interphase, and a colorless upper aqueous phase. RNA remains exclusively in the aqueous phase.

31	32	#
H-miR-25	H-miR-26a	32
M-miR-101b	H-miR-103	64
H-miR-142-5p	H-miR-142-3p	96
H-miR-189	H-miR-190	128
M-miR-211	H-miR-212	160
H-miR-302a	H-miR-302a*	192
H-miR-361	H-miR-362	224
H-miR-409-3p	H-miR-410	256
M-miR-463	M-miR-464	288
H-miR-497	H-miR-498	320
H-miR-518e	H-miR-518f*	352
H-miR-544	H-miR-545	384
H-miR-572	H-miR-573	416
H-miR-604	H-miR-605	448
H-miR-636	H-miR-637	480
M-miR-678	M-miR-679	512
M-miR-710	M-miR-711	544
MAC5		576

3. Transfer the upper aqueous phase carefully without disturbing the interphase into fresh RNase-free tube (see Note 4). The volume of the aqueous phase is about 60% of the volume of TRIzol used for homogenization.
4. Precipitate the RNA from the aqueous phase with isopropyl alcohol. Use 0.5 ml of isopropyl alcohol per 1 ml of TRIzol reagent used for the initial homogenization. Incubate samples at –20°C overnight to precipitate the total RNA including the small RNAs.
5. Centrifuge the precipitates for 15 min at 12,000 × *g* at 4°C (see Note 5).
6. Remove the supernatant. Wash the RNA pellet once with 1 ml of 80% cold ethanol. Centrifuge at 7,500 × *g* for 5 min at 4°C and remove all leftover ethanol carefully.
7. Air-dry or vacuum-dry RNA pellet for 3–5 min (see Note 6). Dissolve RNA in RNase-free water by passing solution a few times through a pipette tip.

8. Measure RNA concentration using Nanodrop. Prepare 50–100 μg of total RNA for array and 20–25 μg for Northern blot in formamide loading buffer (see Note 7) and store at −80°C for future use.

3.3. Small RNA Gel Purification

1. Prepare 15% denaturing polyacrylamide sequencing gel as follows (see Note 8): 60 ml of 30% acrylamide concentrate, 30 ml diluent, 10 ml 10× buffer, 0.04 ml TEMED, and 0.8 ml 10% APS. We recommend using a 1.5 mm spacer, and a comb appropriate for the sample volume.
2. Warm the gel to 45–50°C by pre-running it for 20 min at constant 20 W power in 0.5× TBE running buffer (see Note 9).
3. Mix 50–100 μg of total RNA (see Note 10) with 5 μl external control siRNAs (see Note 11) and then with an equal volume of 2× formamide loading buffer. Heat the samples in formamide loading buffer for 3 min at 95°C.
4. Load RNA samples and run the gel at a constant power of 20–25 W for about 1 h 40 min until the bromophenol blue dye (the leading dye) migrates 2/3 of the gel.
5. Stop the electrophoresis and remove one of the glass plates. Wrap the gel which is still attached to the other glass plate using Saran wrap. Excise a gel band (i.e., the areas marked with samples 1–5 in Fig. 1) between the xylene cyanol marker and the bromophenol blue marker according to the calculations shown in Fig. 1. Cut the gel band into small pieces (~1 mm square). Place the gel pieces in one or two of 2 ml microcentrifuge tubes.
6. Recover the small RNAs by soaking the gel slices in 1 ml of gel elution buffer per tube overnight with rocking at room temperature.
7. Transfer the elution solution that contains the small RNAs into a fresh 2 ml tube and centrifuge for 5 min at 5,000 ×g to remove the residual gel pieces.
8. Add 1 μl glycogen (20 mg/ml) per ml of elution solution and mix thoroughly. Precipitate the small RNAs by adding three volumes of chilled absolute ethanol, mix well and leave overnight at −20°C.
9. Centrifuge the precipitated small RNAs for 15 min at 12,000 ×g at 4°C and remove the supernatant (see Note 12).
10. Wash the small RNA pellet with 1 ml of 80% cold ethanol. Centrifuge at 7,500 ×g for 5 min at 4°C and remove all leftover ethanol carefully.
11. Air-dry or vacuum-dry RNA pellet for 1–3 min.
12. Resuspend the purified small RNAs of each sample in 40 μl RNase-free water.

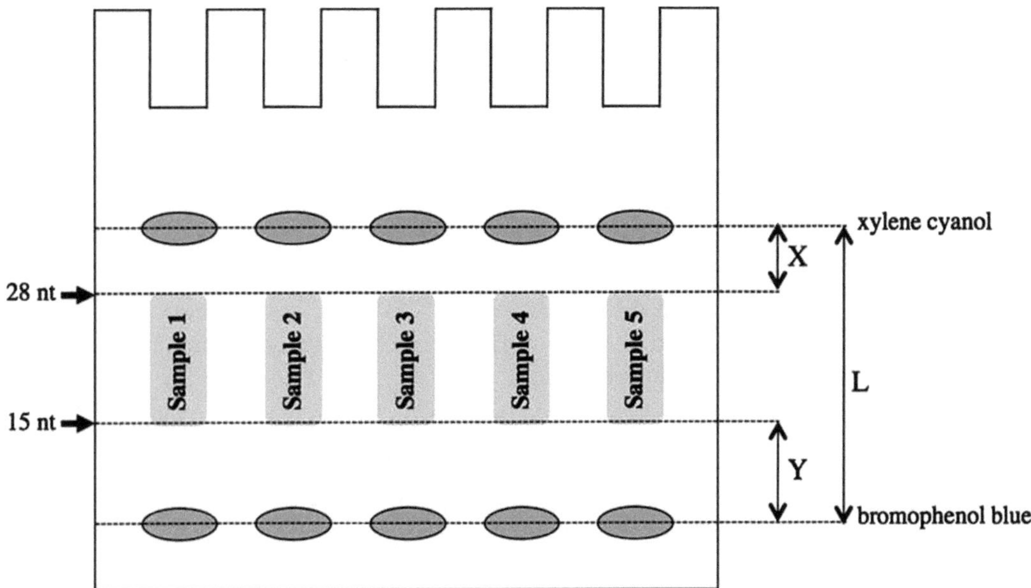

Fig. 1. The electrophoresis position pattern of the 15–28 nt small RNAs. *L* represents the distance between the xylene cyanol marker and the bromophenol blue marker. *X* represents the distance between the xylene cyanol marker and the 28 nt small RNA. *Y* represents the distance between the bromophenol blue marker and the 15 nt small RNA. In a typical PAGE electrophoresis experiment using radiolabeled 15 and 28 nt small RNAs under the conditions that a stable 20 W was applied for 1 h 40 min, we have determined that the L is 4.9 cm, X is 0.6 cm, and Y is 2.2 cm. In each electrophoresis, the *L* can be determined with slight variations. With the known *L* value, *X* and *Y* can be determined by the formula: $4.9/L = 0.6/X = 2.2/Y$. Thus, the area of the gel band containing 15–28 nt small RNAs can be determined by the calculated values of *X* and *Y*.

3.4. Small RNA 5′ Phosphate Removing and Labeling

The bona fide miRNAs and endogenous siRNAs contain a monophosphate at their 5′ ends. So, the 5′-phosphate should be removed (e.g., with Antarctic phosphatase) prior to labeling with polynucleotide kinase (PNK) and γ-^{32}P-ATP.

1. Combine the following dephosphorylation reaction in an RNase-free microfuge tube: 40 μl purified small RNAs, 5 μl Antarctic phosphatase (AP) buffer (10×), 1 μl RNAsin, 1 μl Antarctic phosphatase (5 U), and water to make up the final volume to 50 μl. Incubate at 37°C for 1 h.
2. Inactivate the phosphatase by incubating at 65°C for 10 min.
3. Label the dephosphorylated small RNAs by combining the following: 50 μl small RNA sample, 6.5 μl 10× PNK buffer, 2 μl Polynucleotide Kinase (20 U), 1 μl RNAsin (40 U), 2 μl γ-^{32}P-ATP (6,000 Ci/mmol and 150 mCi/ml), and water to make up the final volume to 65 μl. Incubate at 37°C for 2 h (see Note 13).
4. Purify the reaction products by passing through G25 spin-columns to remove the unincorporated free nucleotides. The purified labeled small RNA is now ready for further hybridization with the array membrane (see Note 14).

3.5. Array Prehybridization and Hybridization

1. Prepare hybridization buffer for prehybridization (10 ml/per array): 5 ml of 100% formamide, 2.5 ml of 20× SSPE, 1 ml 50× Denharts, 0.5 ml 10% SDS, 80 μl 10 mg/ml Sperm DNA (denatured by heating for 5 min and chilling on ice for 5 min), and 0.92 ml H_2O.
2. Put the array membrane in a hybridization tube and wet it with hybridization buffer. Rotate in hybridization chamber at 37°C for at least 2 h.
3. Add the labeled small RNAs to the hybridization buffer in the tube without a direct touch of the membrane after 2 h of prehybridization.
4. Continue the hybridization for 16–24 h at 37°C.
5. Remove the hybridization solution and wash the membrane in 30 ml of pre-warmed washing buffer (2× SSC and 0.1% SDS) for 20 min at 37°C.
6. Repeat washing 2–3 times.
7. Wrap the membrane in Saran wrap and place it in a PhosphorImager cassette for overnight exposure or longer to a PhosphorImager screen (see Note 15).

3.6. Array Image Documentation and Data Analysis

1. Scan the PhosphorImager screen on a PhosphorImager scanner with a scan resolution of 100 μm. An example of representative array images of four types of human prostate cancer cells is shown in Fig. 2.
2. Quantify the image of miRNA array by ImageQuant software and average the radio-signal intensities of duplicate spots for each miRNA.
3. Data normalization. The obtained array data are normalized to external and internal controls to remove system-related variations (e.g., RNA amount variations, small RNA loss during the process, and variations in labeling signal strength). The internal control is a ubiquitously expressed miRNA (such as using let-7 if it is not variable in all the analyzed cancer cell lines) and its expression level in different samples can be obtained by Northern blots (see Subheading 3.8).
4. Clustering analysis. Clustering analysis is performed using Cluster 3.0 with an average linkage and Euclidean distance metric (48). The clustering data were visualized using Java TreeView (49) (Fig. 3).

3.7. Array Membrane Stripping

After image exposure and capture, the array membrane should be stripped for future use.

1. Wash membrane for 30 min to 2 h in stripping buffer (0.1× SSC and 0.5% SDS) at 70°C until no radioactivity can be detected on the membrane.

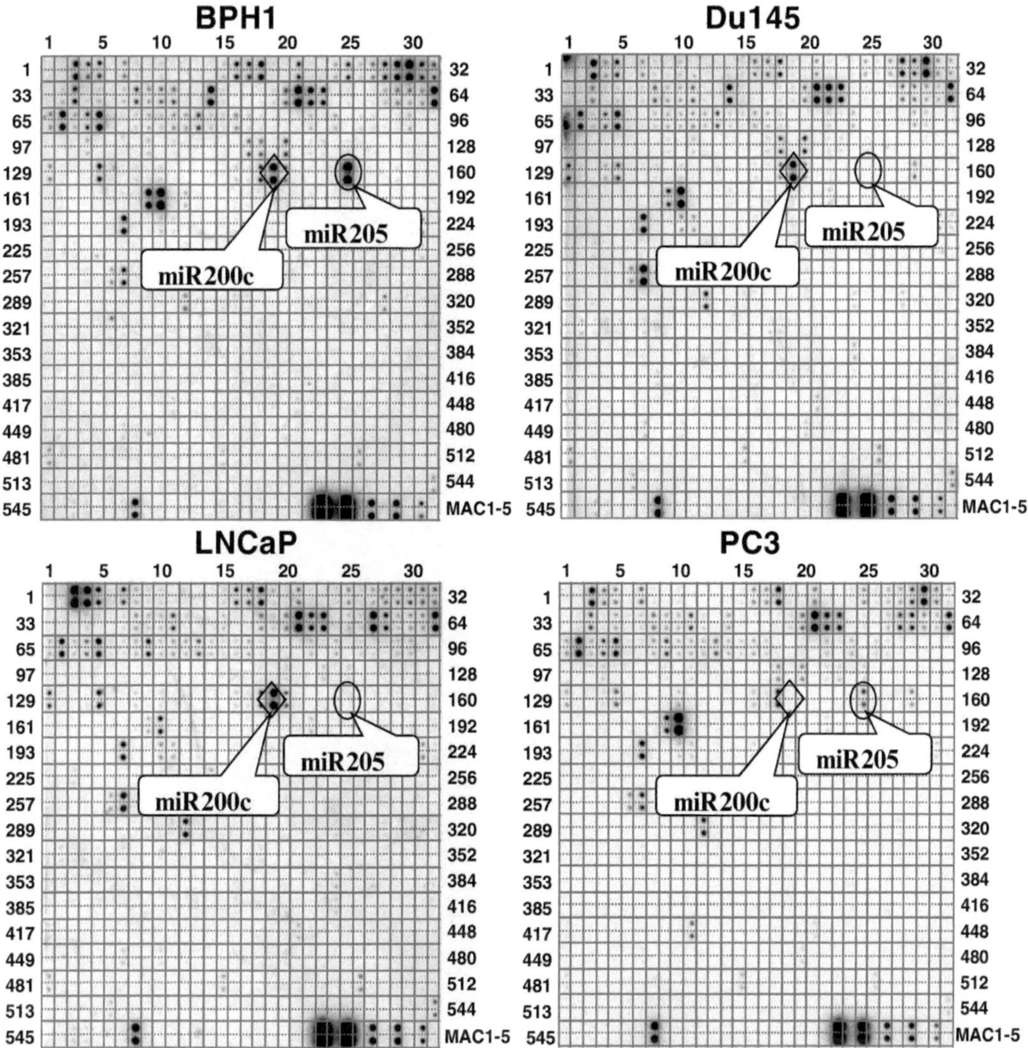

Fig. 2. An array analysis of the miRNA expression profile in different prostate cancer cell lines. Total RNA (100 μg) from BPH-1, DU-145, LNCaP, and PC-3 were separated on 15% PAGE and 15–28 nt small RNAs were isolated from the gel and treated for the array as described in Subheading 3.

2. Check the efficiency of the stripping by exposing the membrane to a phosphorImager screen (see Note 16).
3. The stripped membrane should be wrapped and stored at –20°C in a sealed plastic bag for long-term storage.

3.8. Northern Blot Validation

The differentially expressed miRNAs from different prostate cancer cells should be further validated by Northern blotting analysis. Fig. 4. shows a Northern blot validation of the differentially expressed representative miRNAs. The detailed steps for Northern blot validation are outlined as follows:

1. The total RNAs for Northern blot validation should be from the same batch as for the array analysis (for total RNA isolation, see Subheading 3.2).

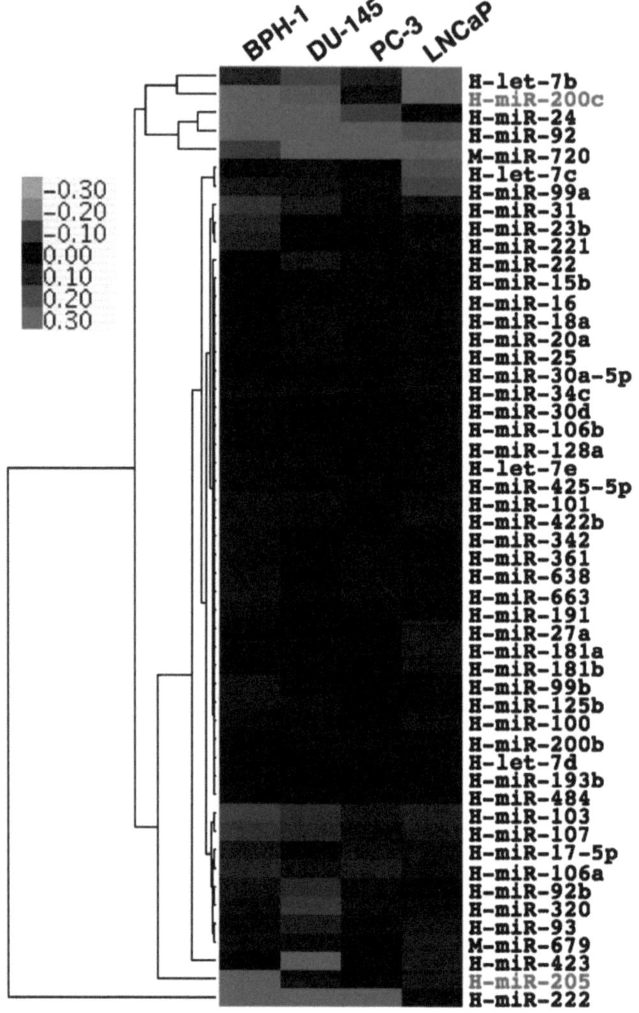

Fig. 3. Hierarchical clustering of miRNAs in different prostate cancer cell lines. Fifty-one miRNAs were clustered based on their expression profiles in BPH-1, DU-145, PC-3, and LNCaP by using Gene Cluster 3.0 (average linkage and Euclidean distance as similarity measure). Data from each miRNA were median centered and normalized. Samples are in columns and miRNAs in rows. The expression values ranged from +0.3 log10 to −0.3 log10.

2. Separate 20–25 μg of total RNA on a gel (see Subheading 3.3) until the bromophenol blue reaches the bottom of the gel.

3. Stain the gel with ethidium bromide (5 μl of 10 mg/ml ethidium bromide in 100 ml of 0.5 X TBE buffer). Visualize the rRNA bands on a UV transilluminator to examine the quality and quantity of the total RNA.

4. Transfer the RNAs to a Hybond N$^+$ membrane by a semi-dry electrotransfer apparatus using 0.5 X TBE buffer for 1 hr at 400 mA.

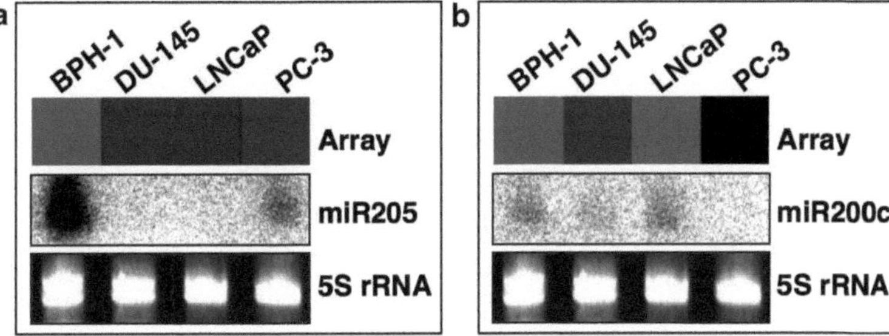

Fig. 4. Northern blot confirmation of the miRNA array data. Total RNA (50 μg per sample) from different prostate cancer cell lines, BPH-1, DU-145, LNCaP and PC-3 were electrophoresed in 15% polyacrylamide gels under denaturing conditions, blotted and hybridized with ^{32}P-labelled miRNA probes. 5S rRNA stained with ethidium bromide was used for loading control. The expression pattern of miR205 (**a**) and miR200c (**b**) were consistent with the array data. The array bars were exported individually from the previous clustering tree.

5. Crosslink the transfered RNAs to the membrane by UV-crosslinker with 1,200 × 100 μJ/cm^2 for 1 min. Store at RT between filter papers until use.
6. Radiolabel miRNA probe: 1 μl of 10 μM antisense DNA probe of target miRNA, 1μl of 10 X T4 PNK buffer, 0.5 μl of T4 PNK, 2 μl γ-^{32}P-ATP (6000 Ci/mmol, 150 mCi/ml) and 5.5 μl of water. Incubate for 1hr at 37°C. Clean up the labeled miRNA probe by passing through G25 spin-column.
7. Prehybridize the membrane in 10 ml of hybridization buffer at 37°C for 1–2 hrs. Add the labeled miRNA probe to the hybridization solution and hybridize overnight at the same temperature.
8. Wash the membrane twice in washing solution (2 × SSC, 0.1% SDS) and expose the membrane on phosphorImager screen and document the radioactive signal (see Subheading 3.6).

3.9. Results and Conclusions

We provide a concise protocol for a simple miRNA array analysis for human prostate cancer cells as an example. Differential expressions of miRNAs have been revealed from four types of human prostate cells at different stages of cancer development (Fig. 2). A total of 51 unique miRNAs were collectively detected in the four prostate cell lines and clustered in Fig. 3. To confirm that the array results are reliable, we validated the expression of miR200c and miR205 that show differential expression in different prostate cell lines using Northern blot techniques. The array and Northern blot validation data are in perfect agreement, which is shown in Fig. 4. These data demonstrate that this simple miRNA array technology together with the related comprehensive protocol provide a reliable, sensitive, and accurate approach to cancer studies. We expect that this array technology and the protocol can be extended to the analysis of miRNAs in all kinds of human diseases.

4. Notes

1. About 1,300 miRNAs have been cloned or predicted from humans and mice (http://www.mirbase.org/). These miRNAs are divided into different families and some family members (paralogs) only differ by a few nucleotides. Thus, cross-hybridization between some miRNAs and their probes is inevitable. Single nucleotide mutation and cross-hybridization experiments showed that miRNAs with a stretch of less than 18 continuous identical nucleotides have no cross-hybridization and specific probes are needed for each of them. On the other hand, miRNAs with a stretch of more than 18 continuous identical nucleotides will have cross-hybridization with their probes and thus only one probe is necessary to detect their additive signals (47). Following this principle and based on the known sequence information of miRNAs from the miRNA database, a total of 553 miRNA antisense DNA oligos were selected to cover all known (up to August 2006) conserved and unique human and mouse miRNAs (4, 5). Since there is no standard internal control to normalize the expression among different treatments, we designed five external control probes (MAC 1-5). The corresponding five synthetic small RNAs complementary to the external controls were added to the total RNA. This helps to monitor the loss of small RNAs during extraction and purification as well as to normalize the array data. Any loss of small RNAs would be represented by the loss of external control signals.
2. RNA integrity is critical for these experiments. Special care should be taken during RNA preparation. All containers and RNA-extracting equipment should be prepared RNase free by DEPC treatment or by baking them overnight. Always wear gloves while handling RNA.
3. An updated miRNA selection should be based on the current miRBase and should also include newly published miRNAs that may not have been entered into the miRBase.
4. While transferring the aqueous phase, make sure that you are not touching the interphase.
5. The RNA precipitate, often invisible after centrifugation, forms a gel-like pellet on the side and bottom of the tube.
6. It is important not to let the RNA pellet dry completely as this will greatly decrease its solubility.
7. RNA is much more stable in formamide.
8. Avoid air bubble formation while casting the gel. Rinse the gel wells thoroughly using the running buffer immediately after pulling out the comb following the gel polymerization.

9. Load the RNA samples onto the wells before the gel temperature cools down.
10. Use the same amount of total RNA for all samples to be arrayed for the purpose of subsequent analysis and comparison.
11. The external control siRNAs is a mixture of different amounts (40, 20, 10, and 5 fmol) of synthetic 21-nt control RNAs that are complementary to the five MAC probes.
12. The precipitated pellets are rather small and glassy. Extreme care is required as the pellet may go with the supernatant while removing supernatant from the tube in which small RNAs are precipitated. It is hard to see the small RNA pellet, if small quantity of RNA is used.
13. All these steps should be done with extreme care and behind a radioactive protective shield. When you are working with radioisotopes, always wear appropriate protective wears such as eye glasses and gloves.
14. Only well-labeled samples should be used for hybridization with the array membrane. Signal from the eluted labeled RNA should be higher than that from the free isotope inside the column. That is, more than 50% isotopes are incorporated into the small RNAs.
15. Never let the membrane dry out until the membrane is stripped. Do not let your membranes stand longer than 2 min without adding the washing buffer. Otherwise, you will not be able to wash away the array background and ruin your array membrane.
16. Abundant miRNAs may require longer stripping treatment. The array membrane can be reused for many times. If you want to reuse the array membrane, never let the membrane dry out. After hybridization, the probe can be stripped and the membrane can be wrapped using a Saran wrap while it is still moist and can be stored at 4°C.

Acknowledgments

X. T. is supported by NIH/NIDDK K01 award (K01DK078648) and R03DK08166. G.T. is supported by the Kentucky Tobacco Research and Development Center (KTRDC), the USDA-NRI grants 2006-35301-17115 and 2006-35100-17433, the NSF MCB-0718029 (Subaward No. S-00000260), and an award from the Kentucky Science and Technology Corporation under Contract # KSTC-144-401-08-029. This work was partially supported by DOD Synergistic Idea Development Awards W81XWH-08-1-0430 (to H.Z.) and W81XWH-08-1-0431 (to N.K.).

References

1. Bartel, D. P. (2004) MicroRNAs: genomics, biogenesis, mechanism, and function. *Cell* **116**, 281–297.
2. Berezikov, E., Cuppen, E., and Plasterk, R. H. (2006) Approaches to microRNA discovery. *Nat Genet* **38 Suppl**, S2–7.
3. Jones-Rhoades, M. W., Bartel, D. P., and Bartel, B. (2006) MicroRNAS and their regulatory roles in plants. *Annu Rev Plant Biol* **57**, 19–53.
4. Griffiths-Jones, S. (2006) miRBase: the microRNA sequence database. *Methods Mol Biol* **342**, 129–138.
5. Griffiths-Jones, S., Saini, H. K., van Dongen, S., and Enright, A. J. (2008) miRBase: tools for microRNA genomics. *Nucleic Acids Res* **36**, D154–158.
6. Bentwich, I. (2005) Prediction and validation of microRNAs and their targets. *FEBS Lett* **579**, 5904–5910.
7. Lewis, B. P., Burge, C. B., and Bartel, D. P. (2005) Conserved seed pairing, often flanked by adenosines, indicates that thousands of human genes are microRNA targets. *Cell* **120**, 15–20.
8. Miranda, K. C., Huynh, T., Tay, Y., Ang, Y. S., Tam, W. L., Thomson, A. M., Lim, B., and Rigoutsos, I. (2006) A pattern-based method for the identification of MicroRNA binding sites and their corresponding heteroduplexes. *Cell* **126**, 1203–1217.
9. Xie, X., Lu, J., Kulbokas, E. J., Golub, T. R., Mootha, V., Lindblad-Toh, K., Lander, E. S., and Kellis, M. (2005) Systematic discovery of regulatory motifs in human promoters and 3' UTRs by comparison of several mammals. *Nature* **434**, 338–345.
10. Calin, G. A., Liu, C. G., Sevignani, C., Ferracin, M., Felli, N., Dumitru, C. D., Shimizu, M., Cimmino, A., Zupo, S., Dono, M., Dell'Aquila, M. L., Alder, H., Rassenti, L., Kipps, T. J., Bullrich, F., Negrini, M., and Croce, C. M. (2004) MicroRNA profiling reveals distinct signatures in B cell chronic lymphocytic leukemias. *Proc Natl Acad Sci USA* **101**, 11755–11760.
11. Hayashita, Y., Osada, H., Tatematsu, Y., Yamada, H., Yanagisawa, K., Tomida, S., Yatabe, Y., Kawahara, K., Sekido, Y., and Takahashi, T. (2005) A polycistronic microRNA cluster, miR-17-92, is overexpressed in human lung cancers and enhances cell proliferation. *Cancer Res* **65**, 9628–9632.
12. Pallante, P., Visone, R., Ferracin, M., Ferraro, A., Berlingieri, M. T., Troncone, G., Chiappetta, G., Liu, C. G., Santoro, M., Negrini, M., Croce, C. M., and Fusco, A. (2006) MicroRNA deregulation in human thyroid papillary carcinomas. *Endocr Relat Cancer* **13**, 497–508.
13. Volinia, S., Calin, G. A., Liu, C. G., Ambs, S., Cimmino, A., Petrocca, F., Visone, R., Iorio, M., Roldo, C., Ferracin, M., Prueitt, R. L., Yanaihara, N., Lanza, G., Scarpa, A., Vecchione, A., Negrini, M., Harris, C. C., and Croce, C. M. (2006) A microRNA expression signature of human solid tumors defines cancer gene targets. *Proc Natl Acad Sci USA* **103**, 2257–2261.
14. Yue, J., and Tigyi, G. (2006) MicroRNA Trafficking and Human Cancer. *Cancer Biol Ther* **5**, 573–578.
15. Calin, G. A., and Croce, C. M. (2006) MicroRNA-Cancer Connection: The Beginning of a New Tale. *Cancer Res* **66**, 7390–7394.
16. Croce, C. M. (2009) Causes and consequences of microRNA dysregulation in cancer. *Nat Rev Genet* **10**, 704–714.
17. Esquela-Kerscher, A., and Slack, F. J. (2006) Oncomirs - microRNAs with a role in cancer. *Nat Rev Cancer* **6**, 259–269.
18. Gregory, R. I., and Shiekhattar, R. (2005) MicroRNA biogenesis and cancer. *Cancer Res* **65**, 3509–3512.
19. McManus, M. T. (2003) MicroRNAs and cancer. *Semin Cancer Biol* **13**, 253–258.
20. Slack, F. J., and Weidhaas, J. B. (2006) MicroRNAs as a potential magic bullet in cancer. *Future Oncol* **2**, 73–82.
21. Iorio, M. V., Casalini, P., Tagliabue, E., Menard, S., and Croce, C. M. (2008) MicroRNA profiling as a tool to understand prognosis, therapy response and resistance in breast cancer. *Eur J Cancer* **44**, 2753–2759.
22. Mattie, M. D., Benz, C. C., Bowers, J., Sensinger, K., Wong, L., Scott, G. K., Fedele, V., Ginzinger, D., Getts, R., and Haqq, C. (2006) Optimized high-throughput microRNA expression profiling provides novel biomarker assessment of clinical prostate and breast cancer biopsies. *Mol Cancer* **5**, 24.
23. Yang, N., Kaur, S., Volinia, S., Greshock, J., Lassus, H., Hasegawa, K., Liang, S., Leminen, A., Deng, S., Smith, L., Johnstone, C. N., Chen, X. M., Liu, C. G., Huang, Q., Katsaros, D., Calin, G. A., Weber, B. L., Butzow, R., Croce, C. M., Coukos, G., and Zhang, L. (2008) MicroRNA microarray identifies Let-7i as a novel biomarker and therapeutic target in human epithelial ovarian cancer. *Cancer Res* **68**, 10307–10314.

24. Iorio, M. V., Ferracin, M., Liu, C. G., Veronese, A., Spizzo, R., Sabbioni, S., Magri, E., Pedriali, M., Fabbri, M., Campiglio, M., Menard, S., Palazzo, J. P., Rosenberg, A., Musiani, P., Volinia, S., Nenci, I., Calin, G. A., Querzoli, P., Negrini, M., and Croce, C. M. (2005) MicroRNA gene expression deregulation in human breast cancer. *Cancer Res* **65**, 7065–7070.

25. Ciafre, S. A., Galardi, S., Mangiola, A., Ferracin, M., Liu, C. G., Sabatino, G., Negrini, M., Maira, G., Croce, C. M., and Farace, M. G. (2005) Extensive modulation of a set of microRNAs in primary glioblastoma. *Biochem Biophys Res Commun* **334**, 1351–1358.

26. Murakami, Y., Yasuda, T., Saigo, K., Urashima, T., Toyoda, H., Okanoue, T., and Shimotohno, K. (2005) Comprehensive analysis of microRNA expression patterns in hepatocellular carcinoma and non-tumorous tissues. **25**, 2537–2545.

27. He, H., Jazdzewski, K., Li, W., Liyanarachchi, S., Nagy, R., Volinia, S., Calin, G. A., Liu, C. G., Franssila, K., Suster, S., Kloos, R. T., Croce, C. M., and de la Chapelle, A. (2005) The role of microRNA genes in papillary thyroid carcinoma. *Proc Natl Acad Sci USA* **102**, 19075–19080.

28. Takamizawa, J., Konishi, H., Yanagisawa, K., Tomida, S., Osada, H., Endoh, H., Harano, T., Yatabe, Y., Nagino, M., Nimura, Y., Mitsudomi, T., and Takahashi, T. (2004) Reduced expression of the let-7 microRNAs in human lung cancers in association with shortened postoperative survival. *Cancer Res* **64**, 3753–3756.

29. Yanaihara, N., Caplen, N., Bowman, E., Seike, M., Kumamoto, K., Yi, M., Stephens, R. M., Okamoto, A., Yokota, J., Tanaka, T., Calin, G. A., Liu, C. G., Croce, C. M., and Harris, C. C. (2006) Unique microRNA molecular profiles in lung cancer diagnosis and prognosis. *Cancer Cell* **9**, 189–198.

30. Michael, M. Z., SM, O. C., van Holst Pellekaan, N. G., Young, G. P., and James, R. J. (2003) Reduced accumulation of specific microRNAs in colorectal neoplasia. *Mol Cancer Res* **1**, 882–891.

31. Bandyopadhyay, S., Mitra, R., Maulik, U., and Zhang, M. (2010) Development of the human cancer microRNA network. *Silence* **1**, 6.

32. Greenlee, R. T., Murray, T., Bolden, S., and Wingo, P. A. (2000) Cancer statistics, 2000. *CA Cancer J Clin* **50**, 7–33.

33. Jiang, J., Lee, E. J., Gusev, Y., and Schmittgen, T. D. (2005) Real-time expression profiling of microRNA precursors in human cancer cell lines. *Nucleic Acids Res* **33**, 5394–5403.

34. Leite, K. R., Canavez, J. M., Reis, S. T., Tomiyama, A. H., Piantino, C. B., Sanudo, A., Camara-Lopes, L. H., and Srougi, M. (2010) miRNA analysis of prostate cancer by quantitative real time PCR: Comparison between formalin-fixed paraffin embedded and fresh-frozen tissue. *Urol Oncol.* doi:10.1016/j.urolonc.2009.05.008.

35. Lin, S. L., Chang, D., and Ying, S. Y. (2007) Hyaluronan stimulates transformation of androgen-independent prostate cancer. *Carcinogenesis,* **28**, 310–20.

36. Shi, X. B., Xue, L., Yang, J., Ma, A. H., Zhao, J., Xu, M., Tepper, C. G., Evans, C. P., Kung, H. J., and deVere White, R. W. (2007) An androgen-regulated miRNA suppresses Bak1 expression and induces androgen-independent growth of prostate cancer cells. *Proc Natl Acad Sci USA* **104**, 19983–19988.

37. Ambs, S., Prueitt, R. L., Yi, M., Hudson, R. S., Howe, T. M., Petrocca, F., Wallace, T. A., Liu, C. G., Volinia, S., Calin, G. A., Yfantis, H. G., Stephens, R. M., and Croce, C. M. (2008) Genomic profiling of microRNA and messenger RNA reveals deregulated microRNA expression in prostate cancer. *Cancer Res* **68**, 6162–6170.

38. Lee, K. H., Chen, Y. L., Yeh, S. D., Hsiao, M., Lin, J. T., Goan, Y. G., and Lu, P. J. (2009) MicroRNA-330 acts as tumor suppressor and induces apoptosis of prostate cancer cells through E2F1-mediated suppression of Akt phosphorylation. *Oncogene* **28**, 3360–3370.

39. Li, T., Li, D., Sha, J., Sun, P., and Huang, Y. (2009) MicroRNA-21 directly targets MARCKS and promotes apoptosis resistance and invasion in prostate cancer cells. *Biochem Biophys Res Commun* **383**, 280–285.

40. Ozen, M., Creighton, C. J., Ozdemir, M., and Ittmann, M. (2008) Widespread deregulation of microRNA expression in human prostate cancer. *Oncogene* **27**, 1788–1793.

41. Porkka, K. P., Pfeiffer, M. J., Waltering, K. K., Vessella, R. L., Tammela, T. L., and Visakorpi, T. (2007) MicroRNA expression profiling in prostate cancer. *Cancer Res* **67**, 6130–6135.

42. Ribas, J., Ni, X., Haffner, M., Wentzel, E. A., Salmasi, A. H., Chowdhury, W. H., Kudrolli, T. A., Yegnasubramanian, S., Luo, J., Rodriguez, R., Mendell, J. T., and Lupold, S. E. (2009) miR-21: an androgen receptor-regulated microRNA that promotes hormone-dependent and hormone-independent prostate cancer growth. *Cancer Res* **69**, 7165–7169.

43. Rokhlin, O. W., Scheinker, V. S., Taghiyev, A. F., Bumcrot, D., Glover, R. A., and Cohen, M. B. (2008) MicroRNA-34 mediates AR-dependent p53-induced apoptosis in prostate cancer. *Cancer Biol Ther* **7**, 1288–1296.

44. Spahn, M., Kneitz, S., Scholz, C. J., Nico, S., Rudiger, T., Strobel, P., Riedmiller, H., and Kneitz, B. (2010) Expression of microRNA-221 is progressively reduced in aggressive prostate cancer and metastasis and predicts clinical recurrence. *Int J Cancer*. doi: 10.1002/ijc.24715

45. Sun, T., Wang, Q., Balk, S., Brown, M., Lee, G. S., and Kantoff, P. (2009) The role of microRNA-221 and microRNA-222 in androgen-independent prostate cancer cell lines. *Cancer Res* **69**, 3356–3363.

46. Wang, G., Wang, Y., Feng, W., Wang, X., Yang, J. Y., Zhao, Y., and Liu, Y. (2008) Transcription factor and microRNA regulation in androgen-dependent and -independent prostate cancer cells. *BMC Genomics* **9 Suppl 2**, S22.

47. Tang, X., Gal, J., Zhuang, X., Wang, W., Zhu, H., and Tang, G. (2007) A simple array platform for microRNA analysis and its application in mouse tissues. *RNA* **13**, 1803–1822.

48. de Hoon, M. J. L., Imoto, S., Nolan, J., and Miyano, S. (2004) Open source clustering software. *Bioinformatics* **20**, 1453–1454.

49. Saldanha, A. J. (2004) Java Treeview--extensible visualization of microarray data. *Bioinformatics* **20**, 3246–3248.

Chapter 7

MicroRNA Knock Down by Cholesterol-Conjugated Antisense Oligos in Mouse Organ Culture

Sharon Kredo-Russo and Eran Hornstein

Abstract

Here, we detail a protocol to design and introduce sequence-specific cholesterol-conjugated antisense oligonucleotides into mouse organ culture. We review design principles for "antagomirs", antisense oligos with a cholesterol-moiety modification at the 3′, and present an optimized method to apply them onto 3D cultured embryonic pancreas. The method offers an approach to study the developmental functions of individual miRNAs and to evaluate miRNA targets, which is significantly faster and simpler than comparable genetics-based approaches.

Key words: Antagomir, miRNA silencing, Cholesterol conjugate, Pancreas, miRNA targets

1. Introduction

Through pairing with partially complementary sites on target mRNAs, microRNAs (miRNAs) mediate posttranscriptional silencing of most of the protein-coding genes in mammalian genomes (1). miRNAs have been implicated in critical processes including differentiation, apoptosis, proliferation, and the maintenance of cell and tissue identity; furthermore, their misexpression has been linked to cancer and other diseases (2).

The function of most miRNA genes is still awaiting gain- and loss-of-function studies. Unfortunately, creating genetic knockouts to determine miRNA function through homologous recombination in murine models is difficult. Additionally, paralogous miRNA genes are often expressed from multiple genomic loci, posing redundant function that further hampers knock-out studies. Thus, a method for effective inhibition of miRNA paralogs is needed.

Tamas Dalmay (ed.), *MicroRNAs in Development: Methods and Protocols*, Methods in Molecular Biology, vol. 732,
DOI 10.1007/978-1-61779-083-6_7, © Springer Science+Business Media, LLC 2011

Presently, loss-of-function studies rely on introduction of chemically modified antisense oligonucleotides, which act as competitive inhibitors of miRNAs. To increase the stability of the antisense oligo backbone, a methyl group is added onto the 2′ O position on the nucleic acid (3). Similarly, locked nucleic acid modifications are also commonly used.

Typically, oligonucleotide inhibitors are introduced into cells by transient transfection, however, when 3-dimensional (3D) tissues are considered for knockdown approaches, the standard transfection protocols become limiting, since the transfected antisense oligo cannot penetrate into the inner layers of the 3D organ.

An effective miRNA inhibitor that has been demonstrated to spontaneously spread into organs, even in the context of whole organisms, is based on a cholesterol conjugate. This was originally developed by the Tuschl and Stoffel laboratories in collaboration with Rajewsky and "Alnylam Pharmaceuticals Inc." and was dubbed "antagomir" (4).

Antagomirs are RNA-like oligonucleotides that harbor the following modifications: (1) complete 2-*O*′-methylation, (2) phosphorothioate backbone, and (3) a cholesterol-moiety at the 3′-end. Antagomirs have been shown to repress miRNA function in several mouse models in vivo (5, 6). Thus, antagomirs hold promises to be effective tissue-penetrating anti-miRNA effectors.

Here, the principle design of antagomirs is given, as was described by Stoffel and coworkers (7). Then, a protocol for the efficient delivery of cholesterol-conjugated antisense oligos into cultured embryonic organs is provided. This protocol makes it possible to examine the effect of miRNA loss-of-function during development and differentiation of isolated organs in a culture setup.

2. Materials

2.1. Organ Culture

1. Dissecting instruments: scissors, straight, 11 cm. Sharp tweezers: Jeweler forceps, 11 cm, and a microsurgery spoon (Apiary Medical, West Milford, NJ).
2. Dissection stereoscope with a light source.
3. Tissue culture incubator at 37°C with 5% CO_2.
4. 70% Ethanol: add 15 ml double-distilled water to 35 ml 100% ethanol.
5. Phosphate buffered saline (PBS): Prepare 10× stock with 1.37 M NaCl, 27 mM, KCl, 100 mM Na_2HPO_4, and 18 mM KH_2PO_4. Autoclave and store at room temperature. Prepare working solution by dilution of one part with nine parts of double-distilled water.

6. Culture medium: Earle's Salts Base M-199 E (SAFC Biosciences, Lenexa, Kansas, USA) supplemented with 10% fetal bovine serum, and 1% penicillin and streptomycin.
7. Non-sterile 10-cm plastic plates (NUNC, Naperville, IL).
8. Sterile 6-cm tissue culture plates (NUNC, Naperville, IL).
9. Timed-pregnant female mice are either purchased from a local provider or mated in-house. Consider an average litter of about eight embryos per female and the following experimental standard (see Note 1): Each experimental condition should be performed in triplicate; thus, (1) no treatment, (2) specific antagomir, (3) an irrelevant antagomir, and (4) a shuffled sequence antagomir make 12 pancreata that are recovered from two pregnant females. As timed-pregnancies are sometimes false-detected, it may be worthwhile ordering three females.

2.2. Immuno-flourescence Staining

1. Paraformaldehyde (PFA, Sigma) at a 20% concentration (w/v in PBS) is aliquoted and kept at –20°C. Freshly dilute the stock 1:5 in PBS to prepare 4% PFA. Note that the solution may need to be carefully heated to dissolve, and then cooled to room temperature for use.
2. Permeabilization solution (PBST): Triton X-100 (Sigma), 1% (v/v) in PBS.
3. Blocking solution: CAS-BLOCK (Zymed laboratories, San Francisco, CA).
4. A primary antibody relevant to the specific experiment, preferentially of a protein whose mRNA is an established target of the miRNA under investigation.
5. A secondary antibody for immunohistochemistry or immunofluorescence (Jackson, West Grove, PA).
6. Nuclear stain: Hoechst 33258 (Sigma) 10 mg/ml in distilled water. The working dilution is 1:10,000.
7. Mounting medium: Immu-mount (Thermo Scientific, Northumberland, UK).
8. Microscope cover slips 24×60 mm, (Menzel-Glazer, Braunschweig, Germany).

3. Methods

3.1. Design and Order Cholesterol-Conjugated Antisense Oligos

1. Retrieve the sequence of the miRNA of interest from miRBase (http://www.mirbase.org/).
2. Obtain the reverse complementary sequence. One web-based way to get the sequence is by following the URL below: http://www.bioinformatics.org/JaMBW//2/1/index.html.

3. Place a standard order for a modified RNA oligo at the website of Dharmacon:

   Surf onto the Dharmacon web site and paste the miRNA reverse complementary sequence into the box in: http://www.dharmacon.com/rna/rna.aspx. Require that all the bases of the oligo are 2′-*O*-methylated. A cholesterol conjugate should be added at the 3′ end of the oligonucleotides (marked as Chol). A phosphorothioate backbone is required in the first two nucleotides at the 5′ end as well as the four last nucleotides at the 3′ end, making a total of six phosphorothioate nucleotides per oligo (marked as ps). The oligo should undergo deprotection, desalting, and HPLC purification, following a consumer request from the manufacturer or following ref. 7. Similar products can be purchased from IDT: http://www.idtdna.com/order/oldorder.aspx?type=RNA or other vencors.

4. Upon arrival, spin the tube down to pool the nucleic acid powder at the bottom of the tube. Prepare the antagomir stock solution at a final concentration of 100 μM, by adding RNAse-free double-distilled water to the lyophilized tube and incubate on ice for 1 h. Aliquot and keep at –80°C.

3.2. Preparation of Pancreas Explants for Culture

1. Timed-pregnant female mice at embryonic day 12.5 (E12.5) are sacrificed and embryos are collected.
2. The embryo-containing embryonic sac and decidua are dissected out, washed in ice-cold PBS, and placed in 10-cm plates containing PBS on ice.
3. The embryo is laid on the bottom of the dish with its right side facing up, as shown in Fig. 1a. Using tweezers carefully cut and remove the limb buds and lateral body wall. This will allow exposure of inner organs.

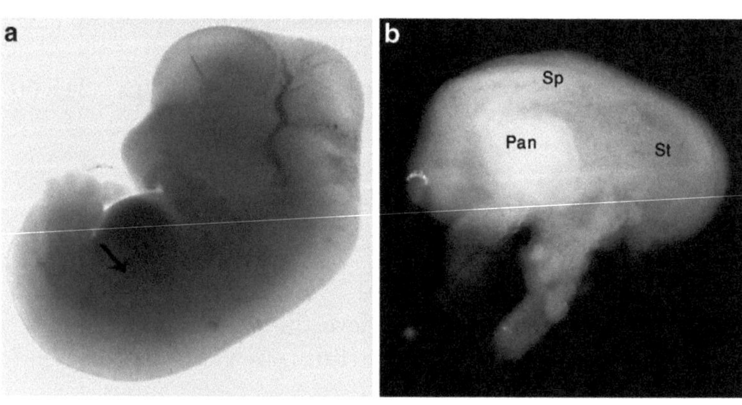

Fig. 1. Pancreas explants are dissected from E12.5 embryos. (**a**) The embryo is laid such that the right side is facing up, with the arrow showing the location of the pancreas. (**b**) High magnification of the stomach (St), spleen (Sp) and dorsal pancreatic bud (Pan).

4. Under the stereoscope, locate the stomach and the spleen.The dorsal pancreatic bud is attached lateral to the stomach (Fig. 1b). Using sharp tweezers, the dorsal pancreas bud is identified and dissected away from adjacent structures such as the stomach and the spleen.
5. The isolated pancreatic bud is gently transferred, using a 10-μl tip, into a 1.5-ml microtube containing cold M-199 medium. Repeat these steps until all pancreata are isolated.
6. The buds are placed for recovery into a 3-cm plate, containing warm M-199 medium and placed at a 37°C, 5% CO_2 incubator, for 1 h.

3.3. Applying the Antagomirs to Hanging Drop Organ Culture

1. The pancreata are next prepared for incubation with antagomir solution in a hanging drop setup (see Fig. 2 for experimental overview). Dilute the antagomir stock to a final concentration of 1 μM in the medium (see Note 2). 30 μl of M-199 medium is needed for each organ submerged in a hanging drop. Each experimental condition should be performed in triplicate.
2. Use 6-cm tissue culture plates. To maintain humidity of the hanging drops culture, fill the bottom of the plate with medium.
3. In order for the surface tension to maintain the drop, make sure that the plastic cover is completely dry.
4. Create the drops by pipetting 30 μl of medium + antagomir on the cover of the 6-cm tissue culture plate. Under the stereoscope, transfer a single explant along with 2 μl of medium and submerge it in an individual drop.
5. Carefully and slowly invert the cover and put it back to its place. Place in a 37°C incubator.
6. Incubate the pancreata for 24–96 h. Under these conditions, the organ recapitulates its in utero development.

After applying antagomirs onto the organ culture, miRNA inhibition should result in derepression of target genes. The expression of target genes is often studied at the protein level, utilizing immunodetection techniques. However, because miRNAs often destabilize the transcript, mRNA quantification is a useful tool as well. See Notes 3 and 4, and for more details see ref. (8).

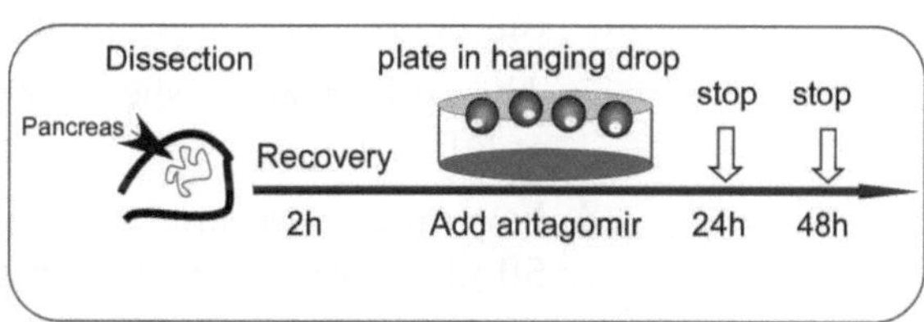

Fig. 2. Experimental overview: at E12.5, the dorsal pancreatic bud is dissected and then cultured in a hanging drop setting with antagomir and the organ phenotype is examined after 24–96 h.

3.4. Immuno-fluorescence Staining: Fixation, Primary and Secondary Antibody, and Image Acquisition

3.4.1. Fixation

1. After 24–96 h the explants are removed from the tissue culture incubator. The cover harboring the drops that maintain the cultured pancreata in drops is inverted and the explants are picked under the stereoscope into a 1.5-ml microtube, prefilled with room temperature (RT) PBS. Tubes are then gently agitated for 5 min at RT (see Note 5).
2. The PBS is discarded and explants are fixed in 4% fresh PFA for 40 min at RT with gentle agitation. Note that PFA is highly toxic and its handling should be conducted in a fume hood.
3. The PFA is discarded and the explants are washed twice with 1 ml PBS for 5 min at RT.
4. The PBS is discarded and the explants are submerged in 1 ml of 70% ethanol.
5. At this point, explants may be stored at 4° until staining. For long-term storage, prevent possible evaporation by sealing the microtube lids with parafilm.

3.4.2. Immunoflourescence

1. Graded rehydration from 70% ethanol into PBS is done by incubation of the explants in 50% ethanol/PBS for 1 h at RT on a rocking platform. Next, explants are washed twice with 1 ml PBS for 10 min each at RT.
2. The PBS is removed and replaced with 1 ml permeabilization solution (PBST) for 1 h at RT.
3. The explants are then incubated in 150 μl blocking buffer for 2 h (or more) at RT on a rocking platform.
4. The primary antibody is prepared in blocking solution +0.5% Triton X-100, and 100 μl is added to the explants for overnight incubation at 4°C on a rocking platform (see Note 6).
5. The next day, explants are washed three times with 1 ml PBST 1% for 10 min each at RT with agitation. Note that the primary antibody solution may be kept for reuse in the future, if kept chilled or frozen.
6. The secondary antibody is freshly prepared for each explant at a 1:200 dilution in blocking buffer. The explants are incubated in 100 μl of the secondary antibody solution for 3–4 h at RT on a rocking platform (as an alternative, the explants can be incubated over night at 4°C).
7. The explants are washed tree times with 1 ml PBST for 10 min at RT, shaking.
8. For nuclear counterstaining, dilute Hoechst 1:10,000 in 50 μl of distilled water per explant (1 μg/ml). The explants are then incubated for 1 h at RT with agitation.
9. Finally, the Hoechst solution is discarded and replaced with 1 ml PBS and the explants are washed twice for 10 min at RT.

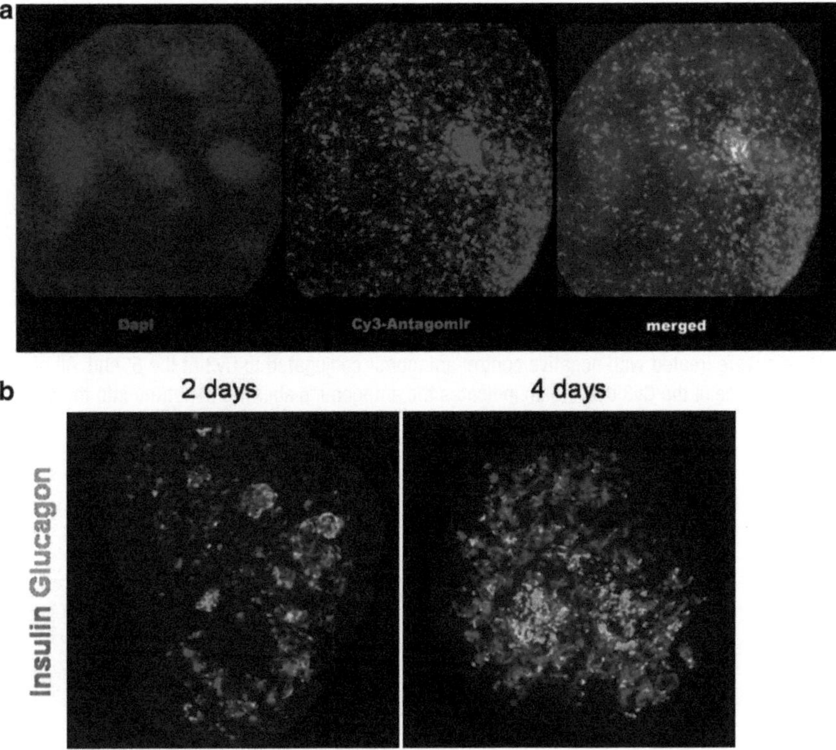

Fig. 3. (**a**) E12.5 pancreas explants were treated with negative control antagomir conjugated to a Cy3 at its 5′ end. Confocal Z-sections spanning 100 μm, demonstrate effective and homogenous penetration of the antagomir through multiple cell layers of the 3D organ, (**b**) IHC staining for Insulin and Glucagon demonstrates pancreas explant differentiation in culture.

10. Explants are mounted on a microscope slide. Then, a drop of mounting medium is placed on top of the tissue, followed by a coverslip. Once the mounting medium is dried, the slides can be viewed or stored in the dark at 4°C.
11. The slides are viewed under standard phase contrast microscopy or confocal microscopy (Fig. 3).

4. Notes

1. Control experiments are required to ensure that the studied effect is sequence-specific. Thus, experiments should employ a sequence-specific antagomir and negative controls. Suggested negative controls are composed of mismatched/shuffled sequence oligos and/or an antagomir for an unrelated miRNA. Further, it is wise to require that one of the above controls is Cy3-conjugated at the 5′ end, allowing evaluation of the antagomir penetration into the organ (see Fig. 4).

Examples of suitable negative controls:

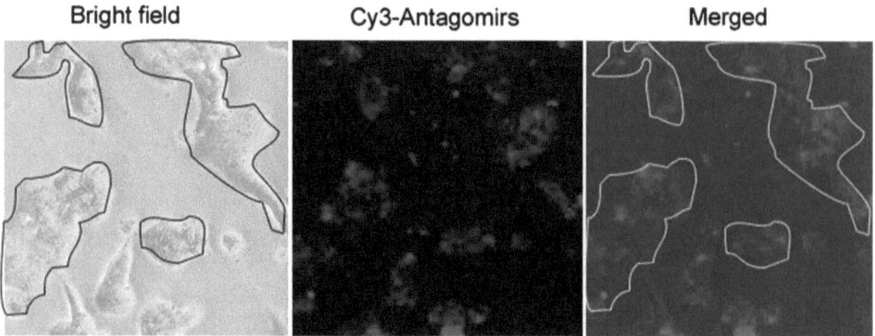

Fig. 4. BetaTC cells were treated with negative control antagomir conjugated to Cy3 at the 5′ end. All of the cells show intracellular fluorescence of the Cy3 dye, which indicates the antagomir's ability to penetrate into the cell.

Table 1
Antagomir sequences

Antag_*C.elegans*	UpsCpsUACUCUUUCUAGGAGGUUGpsUpsGps Aps-Chol
Antag_122	ApsCpsAAACACCAUUGUCACACUpsCpsCpsAps-Chol

(1) Antisense sequence of a nonhomologous gene from another species, for example antagomir to cel-miR-67 from *Caenorhabditis elegans* (Table 1).

(2) For most tissues other than the liver, consider using the antisense sequence of the liver-specific miRNA, miR-122, that is less abundant in other tissues (Table 1).

(3) A scrambled version of the experimental antagomir may be based on shuffled sequences. Shuffled sequences for many known miRNAs are available in Table S1 of ref. (9).

2. The antagomir concentration depends on the miRNA expression levels and needs to be determined empirically for each gene studied. We recommend calibrating the dose starting from 50 nM and up to 2 μM, monitoring expression of a known target gene as readout.

3. A suggested preliminary step in evaluation of the antagomir is to introduce it into cultured cells and analyze its effect on derepression of a reporter construct. This approach was used, for example, to show the effect of antagomir against miR-122 in ref. 4. Briefly, the reporter construct should harbor artificial seed-binding sequences or a genuine fragment from an established target, downstream of a luciferase reporter, and thus enables assessment of the antagomir as a repressor the miRNA activity.

4. Evaluation of miRNA repression can also be monitored with quantitative real-time PCR to a previously known target. This can be achieved by standard real-time PCR protocol that measure gene expression levels as in ref. (10).
5. All washes and staining of the explants are done in a standard 1.5-ml microtubes, unless stated otherwise.
6. The primary antibody dilution depends on the antibody being used. We often used GP-Insulin (1:200) and Rb-Glucagon (1:300), both from Dako Cytomation, Glostrup, Denmark.

Acknowledgments

We thank Judith Maggenheim and Yuval Dor (Hadassah Medical School, The Hebrew University, Jerusalem, Israel) for protocols and advice and Tal Melkman-Zehavi for comments on this manuscript. EH is the incumbent of the Helen and Milton A Kimmelman Career Development Chair. This work was supported by grants from Juvenile Diabetes Research Foundation (#99-2007-71), the EFSD/D-Cure Young Investigator award, the Israel Science Foundation, the Yeda-Sela Center for Basic Research and the Wolfson Family Charitable Trust.

References

1. Lewis BP, Burge CB, Bartel DP. (2005). Conserved seed pairing, often flanked by adenosines, indicates that thousands of human genes are microRNA targets. *Cell* **120**(1), 15–20.
2. Bartel, D. P. (2004). MicroRNAs: genomics, biogenesis, mechanism, and function. *Cell* **116**(2), 281–97.
3. Tuschl T, Zamore PD, Lehmann R, Bartel DP, Sharp PA. (1999). Targeted mRNA degradation by double-stranded RNA in vitro. *Genes Dev* **13**(24), 3191–7.
4. Krützfeldt J, Rajewsky N, Braich R, Rajeev KG, Tuschl T, Manoharan M, Stoffel M. (2005). Silencing of microRNAs in vivo with “antagomirs”. *Nature* **438**(7068), 685–9.
5. Morton SU, Scherz PJ, Cordes KR, Ivey KN, Stainier DY, Srivastava D. (2008). microRNA-138 modulates cardiac patterning during embryonic development. *Proc Natl Acad Sci USA* **105**(46), 17830–5.
6. van Solingen C, Seghers L, Bijkerk R, Duijs JM, Roeten MK, van Oeveren-Rietdijk AM, Baelde HJ, Monge M, Vos JB, de Boer HC, Quax PH, Rabelink TJ, van Zonneveld AJ. (2008). Antagomir-Mediated Silencing of Endothelial Cell Specific MicroRNA-126 Impairs Ischemia-Induced Angiogenesis. *J Cell Mol Med.* **13**(8A),1577–85.
7. Krützfeldt J, Kuwajima S, Braich R, Rajeev KG, Pena J, Tuschl T, Manoharan M, Stoffel M. (2007). Specificity, duplex degradation and subcellular localization of antagomirs. *Nucleic Acids Res* **35**(9), 2885–92.
8. Farh KK, Grimson A, Jan C, Lewis BP, Johnston WK, Lim LP, Burge CB, Bartel DP. (2005). The widespread impact of mammalian MicroRNAs on mRNA repression and evolution. *Science* **310**(5755), 1817–21.
9. Lewis BP, Shih IH, Jones-Rhoades MW, Bartel DP, Burge CB. (2003). Prediction of mammalian microRNA targets. *Cell* **115**(7), 787–98.
10. Nolan T, Hands RE, Bustin SA. (2006). Quantification of mRNA using real-time RT-PCR. *Nat Protoc* **1**(3), 1559–82.

Chapter 8

Protocols for Use of Homologous Recombination Gene Targeting to Produce MicroRNA Mutants in *Drosophila*

Ya-Wen Chen, Ruifen Weng, and Stephen M. Cohen

Abstract

MicroRNAs (miRNAs) are noncoding RNA molecules that have come to attract considerable interest for their roles in animal and plant development and disease. One means to study miRNA function in animal development is to create mutations. Use of gene-targeting strategies based on ends-out homologous recombination is a useful approach to produce mutations of desired structure, and is gaining popularity for producing miRNA knockouts. Here we present a detailed protocol for miRNA gene targeting and for their subsequent molecular characterization as well as confirmation by rescue. The descriptions of a series of modified vectors designed to facilitate the analysis of miRNA function are included, and a method to manipulate the mutant genome using recombinase-mediated cassette exchange.

Key words: MicroRNA, *Drosophila*, Gene targeting, Homologous recombination, *ϕC31* integrase, Recombinase-mediated cassette exchange

1. Introduction

MicroRNAs (miRNAs) are short noncoding RNAs that act as regulators of posttranscriptional gene expression. miRNAs are products of endogenous genes. Hundreds of miRNA genes have been identified in animal, plant, and viral genomes (reviewed in (1–3)). Analysis of miRNA functions has been greatly aided by generation of mutants that remove their expression. The first two animal miRNAs were found in *Caenorhabditis elegans* as heterochronic mutations, regulating developmental progression between larval stages (4, 5). Identification of these loci by forward genetics was possible because genetic screens in *C. elegans* can sample hundreds of thousands of chromosomes. Consequently, the small target-size of the typical miRNA locus does not pose an insurmountable

Tamas Dalmay (ed.), *MicroRNAs in Development: Methods and Protocols*, Methods in Molecular Biology, vol. 732,
DOI 10.1007/978-1-61779-083-6_8, © Springer Science+Business Media, LLC 2011

problem for forward genetics in the worm. This poses a more intractable problem in other genetic systems. miRNA genes were also identified by gain-of-function genetic screens in the fruit fly *Drosophila*, using the EP-system (6). Transposon-mediated insertional mutagenesis provides a different way to circumvent the target-size problem, since the promoters of miRNA genes are typical of DNA Polymerase II regulated genes, and have proven to be good targets for *P*-element-mediated integration. A number of *Drosophila* miRNA genes have been targeted in this way (7–10).

Animals lacking the functions of all miRNA show severe defects at early stages in most animals. In *C. elegans*, *dicer-1* mutants display defects in germ-line development (11, 12). Loss of both maternal and zygotic *dicer-1* leads to embryonic lethality, suggesting essential roles for some miRNAs during *C. elegans* embryogenesis (13). In *Drosophila*, *dicer-1*-mutant germline stem cells display cell division defects (14) and fail to maintain stem cell fate (15). *dicer-1*-mutant zebrafish die by 2–3 weeks of age, showing a general growth arrest (16). In studies using conditional mouse *Dicer*, embryonic stem cells were impaired in their ability to proliferate (17), and those selected for survival failed to differentiate (18). These studies indicate the importance of miRNAs during development and hint at a diversity of functions. Yet, most individual miRNA remains to be studied in depth due to the lack of specific mutations.

As indicated above, there are many ways to generate mutations removing a specific miRNA. Conventional chemical mutagenesis strategies can be effective, but they are inefficient. In *Drosophila*, nearby *P*-element insertions can inactivate the miRNA gene but also provide the means to make flanking deletions removing the miRNA (7–10, 19). While effective, flanking-deletion strategies based on imprecise *P*-element excision or male recombination are dependent on the proximity of the *P*-element to the miRNA (20, 21) and may be unfeasible if the miRNA is close to another gene, or nested in the intron of a protein-coding gene.

The advent of gene targeting by ends-out homologous recombination has provided a direct way to target specific loci, by creating precise user-defined deletions (22–27). This approach has been applied with considerable success to producing miRNA mutants in the fly (28–33).

In this chapter, we describe methods used to make site-specific mutations by homologous recombination in flies, with emphasis on their application to miRNA loci. The ends-out homologous recombination designed by Golic and his colleagues is the basis of this protocol to replace the target miRNA without duplicating the flanking region (22, 25, 26, 34). Figure 1 provides a schematic overview of the steps involved in generating a targeted allele. In brief, a targeting construct is generated by cloning the sequences flanking the gene of interest into an ends-out gene targeting

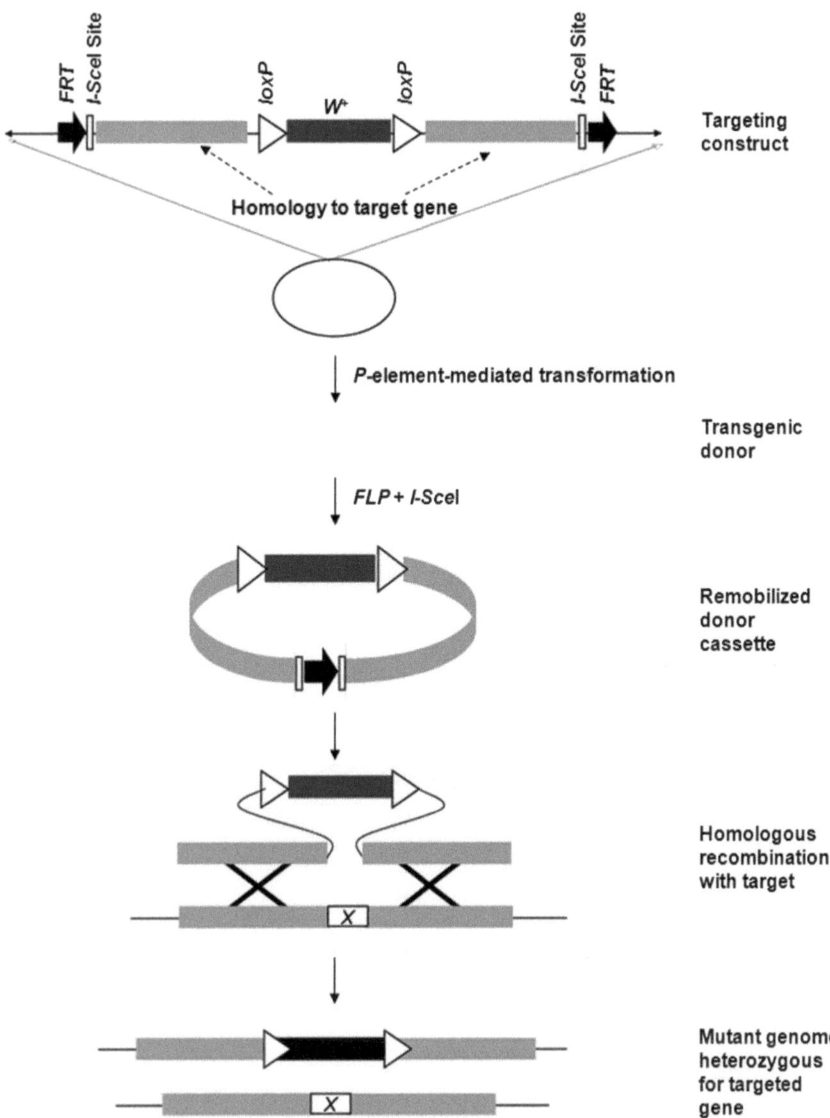

Fig. 1. Schematic representation of ends-out gene targeting by homologous recombination. 3.5–4 kb homology sequences flanking the target gene are cloned into the pW25 ends-out vector. *P*-element-mediated transformation gives rise to transgenic donor containing the targeting cassette. Induction of FLP recombinase is used to excise a circular DNA molecule containing the targeting vector, which is then linearized by cleavage with the I-*Sce*I meganuclease. The linearized targeting vector can then recombine with the chromosomal target locus, replacing the endogenous gene with *mini-white* and generating a mutant heterozygous for the targeted miRNA. Figure reproduced from ref. 37 with permission of the Genetics Society of America.

vector, such as pW25 (26) (Fig. 2a). A transgenic donor line is then established based on *P*-element-mediated transformation (Fig. 1). Upon expression of the *FLP* recombinase (to excise a circular molecule of DNA from the construct) and the I-*Sce*I nuclease (to convert the extrachromosomal circle to a linear molecule), the targeting construct is able to undergo homologous

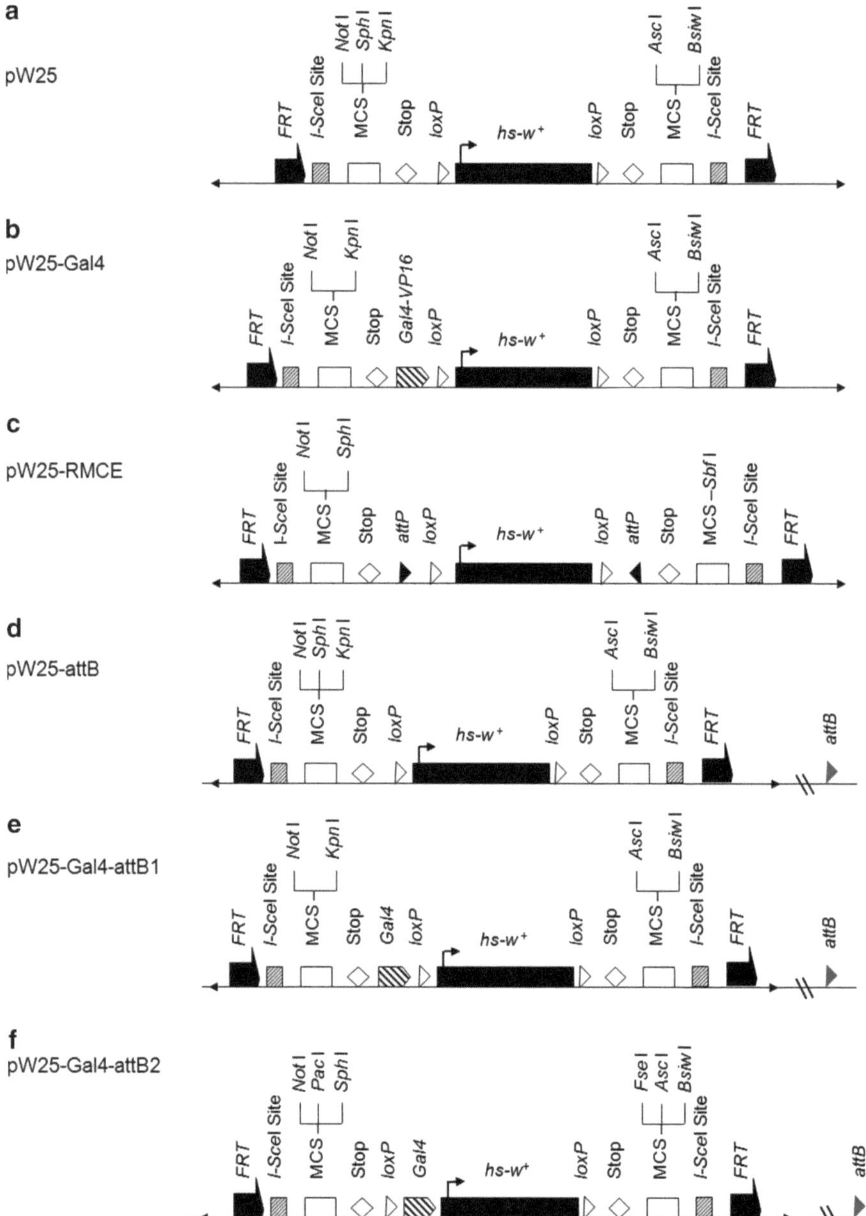

Fig. 2. Modified ends-out gene-targeting vectors. (**a**) pW25 vector (26). *Arrowheads* are *P*-element ends, *half-arrows* are FRT sequences, *hatched boxes* are I-*Sce*I sites, *open boxes* are cloning sites for homologous DNA sequences, *diamonds* are six-frame translation stop codons, and *black boxes* are *mini-white* marker gene. (**b**) pW25-Gal4 vector. The *Gal4* coding sequence, denoted by *hatched pentagon*, is inserted between the 5′ six-frame stop codons and *lox*P site. (**c**) pW25-RMCE vector. Inverted *att*P, denoted by *black triangles* are inserted upstream and downstream of the *mini-white* sequence, respectively. (**d**, **e**) pW25-*att*B and pW25-Gal4-*att*B1 vectors. *att*B, denoted by *grey triangles* was cloned into both pW25 and pW25-Gal4 vector backbones. (**f**) pW25-Gal4-*att*B2 vector. The *Gal4* coding sequence is located between the 5′ *lox*P site and *mini-white* by composite cloning using pW25-*att*B as the vector backbone. This allows removal of *Gal4* driver together with *mini-white* marker using the Cre/*lox*P recombinase system. Restriction enzyme sites for *Pac*I and *Fse*I were introduced to the multiple cloning sites (MCS) at the 5′ and 3′-end, respectively, to facilitate directional cloning of homology arms. Figure adapted from ref. 37 with permission of the Genetics Society of America.

recombination with the endogenous DNA at the target locus. A successful targeting event results in the replacement of endogenous gene of interest by the vector-encoded sequences, including the *mini-white* marker which can be used to identify flies carrying the integrated DNA at the correct chromosomal location. This strategy allows for efficient targeting of any gene of interest. From construct cloning to your first knock-out flies typically takes about 4–5 months. In case things don't go smoothly, troubleshooting hints are provided to highlight the most typical sources of difficulty.

Several groups have reported new generations of targeting vectors, designed to enhance the range of options available (e.g., ref. 35–37). These include introduction of *Gal4* or *GFP* at the targeted locus (knock-in) and use of recombinase-mediated cassette exchange (RMCE) (38) or site-specific integrase-mediated repeated targeting (SIRT) (35) techniques. Our versions of such vectors are described in Fig. 2 (37).

2. Materials

2.1. Fly Stocks

1. Mutant strain for *white* locus, which is useful for detecting transgenes with the *mini-white* reporter by eye color: w^{1118}.
2. Double-balancer stock used for chromosome mapping (e.g., *w**; *Bl/CyO*; *TM2*, *Ubx/TM6B*, *Tb*).
3. Site-specific landing strains used for integrase-mediated transformation: Landing site on chromosome *2* with *ϕC31* integrase on *X* chromosome: y^1, *M{vas-int.Dm}ZH-2A*, *w**; *M{3xP3-RFP.attP'}ZH-22A* (39), landing site on chromosome *2*: y^1, w^{67c23}; $P\{y^{+t7.7}=CaryP\}attP16$ (40), landing site on chromosome *3* with *ϕC31* integrase on chromosome *4*: *y1*, *w**; *M{3xP3-RFP.attP}ZH-86Fb*; *M{vas-int.B}ZH-102D* (39), and landing site on chromosome *3*: y^1, w^{67c23}; $P\{y^{+t7.7}=CaryP\}attP2$ (41).
4. Strains providing expression of FLP recombinase and I-*Sce*I endonuclease: FLP and I-*Sce*I on chromosome *2*: y^1, *w**; $P\{ry^{+t7.2}=70FLP\}11$, $P\{v^{+t1.8}=70I\text{-}SceI\}2B$, noc^{Sco}/CyO, S^2, FLP and I-*Sce*I on chromosome *3*: y^1, *w**; $P\{ry^{+t7.2}=70FLP\}23$, $P\{v^{+t1.8}=70I\text{-}SceI\}4A/TM6$, *Ubx* (22).
5. Strains providing expression of Cre recombinase: Cre on chromosome *3*: *w**; $Kr^{If\text{-}1}$, *Slbo-lacZ(1310)/CyO*; *hs-Crew*$^+$*/TM3*, *Ser*, and Cre on *X* chromosome: *w**, *hs-Crew*$^+$*/FM6*; *Sb/TM3*, *Ser*.
6. Balancer strains: Balancer on *X* chromosome (e.g., y^1, *w**, *lethal/FM6*), on chromosome *2* (e.g., *w**; $Kr^{If\text{-}1}$, *Slbo-lacZ(1310)/CyO*), on chromosome *3* (e.g., *w**; *TM3*, *Sb*, *Ser/TM6B*, *Tb*).

7. *ϕC31* integrase on chromosome *4* which is used for RMCE: *y1*, *M{3xP3-RFP.attP}ZH-2A*, *w**; +; +; *M{eGFP.vas-int.B} ZH-102D* (39).

2.2. Targeting Vector Cloning

1. Phusion High-Fidelity DNA Polymerase (Finnzymes).
2. TOPO XL PCR Cloning Kit (Invitrogen).
3. QiAquick Gel Extraction Kit (Qiagen).

2.3. Verification of Targeting Event in Chromosomal DNA

1. Squashing buffer (SB): 10 mM Tris–glycine, pH 8.2, 1 mM EDTA, 25 mM NaCl, and 200 μg/ml Proteinase K. Proteinase K should be diluted freshly from a frozen stock before use (42).
2. DNeasy Blood & Tissue Kit (Qiagen).

3. Methods

miRNA genes may produce primary transcripts of varying length and complexity. Irrespective of the nature of the primary transcript, or its relation to other nearby genes, the mature miRNA usually derives from a short hairpin of ~80 nucleotides in length. A simple effective design to generate a null mutation for a miRNA involves deleting the entire hairpin (see Note 1). Such short deletions can often be made with high efficiency. Longer deletions of 1 kb or more are possible in cases where several miRNA loci are clustered in a common primary transcript (e.g., miR-309 cluster (31)).

Design of the targeting vector can be quite simple if the miRNA derives from a unique transcription unit that is far from other genes. Additional care should be taken when the sequences to be targeted are close to another gene (even when they are not part of the same transcription unit). Sequences homologous to the arms of the targeting vector may be replaced in the genome. It is therefore important to ensure that the targeting vector is free of sequence changes relative to the host strain (see Note 2). Many miRNAs are located in introns of protein-coding genes (on one strand or both). In such cases, specific strategies are needed to ensure that only the miRNA and not the host gene is mutated.

3.1. Vectors for Gene Targeting

A range of targeting vectors have been produced to add functionality beyond the straightforward gene replacement strategy incorporated into the design of the original pW25 vector (26). A set of vectors suitable for different purposes is illustrated in Fig. 2 (37).

1. pW25 (Fig. 2a) uses *mini-white* as a convenient eye-color marker for both the transformation and targeting processes. DNA flanking the sequences to be deleted are cloned into the

two multiple cloning sites (MCS) to produce the homology arms of the vector. The targeted sequences between these arms will be replaced by the *mini-white* gene, which can be used as a dominant genetic marker for tracking the targeted allele. If desired, *mini-white* can subsequently be removed using Cre recombinase acting on the flanking *lox*P sites (43). After removal of the *mini-white* marker the targeted site contains a single *lox*P site and the two stop cassettes in place of the deleted DNA (see Note 3, illustrated in Fig. 5a). pW25 is a good choice for the generation of simple gene deletions.

2. pW25-Gal4 (Fig. 2b): This modification introduces *Gal4-VP16* upstream of the 5′ *lox*P site in pW25. pW25-Gal4 incorporates the Gal4/UAS binary system into the standard ends-out targeting strategy. Targeting based on this vector can be used to generate a mutant allele that drives *Gal4* expression under control of the endogenous regulatory elements at the targeted locus. Unlike the *mini-white* marker, inserted *Gal4* cannot be removed using Cre recombinase. This produces a "knock-in" mutant with *Gal4-VP16* in place of the deleted sequences. Such alleles can be useful for restoring expression of any desired gene under spatial and temporal control of the miRNA gene. This may be useful, for example, to rescue a mutant by restoring expression of the miRNA.
3. pW25-RMCE (Fig. 2c): This vector was modified to incorporate the recombinase-mediated cassette exchange (RMCE) strategy (38). An acceptor cassette was created in pW25 by flanking *mini-white* with inverted *att*P sites (37) (see Note 4). Targeting with this vector introduces *mini-white* flanked by inverted *att*P sites (as well as by *lox*P sites). Exchange of the chromosomal acceptor cassette with a plasmid-borne donor cassette can be catalyzed with high efficiency by the bacteriophage *ϕC31* integrase (illustrated in Fig. 6). This allows replacement of *mini-white* with any desired sequence. This vector permits replacement of the targeted sequences with any sequence flanked by *att*B sites and can be used for sequential targeting of the same locus with a variety of sequence variants. This may be particularly useful for site-directed mutagenesis to explore protein-coding sequence variants, for rescuing a mutant by restoring expression of the miRNA, or for manipulating miRNS target expression levels in the endogenous miRNA expression domain.
4. pW25-*att*B (Fig. 2d) represents a minor modification of pW25. Introduction of a single bacteriophage attachment site, *att*B, at the *Nde*I site of the vector backbone allows site-specific targeting of the vector to a defined chromosomally located *att*P site during the initial establishment of the transgenic donor line (37). Use of bacteriophage *ϕC31* integrase-mediated transformation

can improve the efficiency of generating the transgenic stain carrying the targeting vector, compared to conventional *P*-element-mediated transformation (39–41). Furthermore, it ensures that the donor transgene is located on the desired chromosome, which eliminates two generations of crosses that would otherwise be needed to map donor transgenes generated by conventional *P*-element integration strategies to the desired chromosome.

5. pW25-Gal4-*att*B1 (Fig. 2e) is similar to pW25-*att*B, with the addition of the *Gal4-VP16* coding sequence outside the *lox*P cassette (as in pW25-Gal4). This combines the improved efficiency of donor transgene production with allowing *Gal4* knock-in at the targeted locus.
6. pW25-Gal4-*att*B2 (Fig. 2f) is similar to pW25-Gal4-*att*B1 except that the *Gal4-VP16* coding sequence was inserted between the 5′ *lox*P site and the *mini-white* gene (see Note 5). *Gal4* and *mini-white* can both be removed using the Cre recombinase, resulting in a "clean" knockout.

3.2. Preparing the Targeting Vector

1. The first step is to select homology arms to be amplified and cloned into the targeting vectors (see Fig. 1). We prefer to use homology arms of 3.5–4 kb when possible. Arms as short as 2.5 kb have been used, but this may come at some cost of overall efficiency (see Note 6). Primer design for PCR amplification of the homology arms is important (see Note 7). There are few unique cloning sites available on the vectors (Fig. 2). *Not*I is most commonly used site for upstream arm cloning, and *Asc*I is most commonly used for the downstream arm (see Note 8). These sites can be incorporated into the primers used for amplification of the homology arms. Some vectors allow directional cloning of the arms using different sites.
2. DNA for the homology arms is amplified from wild-type genomic DNA by PCR. We have had good experience with Phusion High-Fidelity DNA Polymerase to minimize errors introduced during PCR amplification. Other suitable choices include TaKaRa LA *Taq* or Roche High-Fidelity *Taq* (see Note 9). It is important to bear in mind that PCR amplification can introduce mutations. Strategies such as recombination-based cloning ("recombineering"; (44)) provide an alternative approach.
3. PCR products are run on agarose gels to confirm their size and DNA recovered by standard gel extraction procedures.
4. Cloning of the PCR product may be desirable as an intermediate step in the construction of the vector. Depending on the nature of the enzyme used to amplify the fragment, and the cloning strategy preferred, it may be necessary to add an

A-tail to the product for "TA" cloning. This is not needed if using *Taq*-family enzymes. Phusion DNA Polymerase does not add an A-tail to the PCR product, and so cannot be used directly. To add an A-tail for "TA" cloning, incubate the purified PCR product with dATP and *Taq* DNA Polymerase at 72°C for 15 min.

5. A-tailed PCR fragments can be cloned by "TA" cloning, for example, using the TOPO XL PCR cloning Kit, which is designed for cloning large PCR fragments.
6. We recommend sequencing the homology arms, with particular attention to any protein-coding sequences that they contain, to avoid introducing mutations that may occur during PCR amplification into the neighboring genes. Single nucleotide polymorphisms are often found when comparing the sequences amplified from lab strains with the reference genome. We recommend comparing with sequence of an independent PCR reaction using different batch of wild-type genomic DNA to determine whether these are "real" polymorphisms in the lab strain or if they represent PCR-induced mutations (see Note 10).
7. Standard methods are used to clone the PCR products into the targeting vector (see Subheading 3.1). Remember to determine the orientation of the insert if not using a directional strategy. The order in which the upstream or downstream arms are inserted usually does not matter, unless a site for the one of the enzymes is present. In practice, we have found one order may work better than the other, so be prepared to try both if you run into difficulty.
8. We recommend sequencing the ends of both arms to confirm their orientation.
9. Prepare DNA for microinjection/transgene production using standard methods.

3.3. Generation of Transgenic Donor Lines

Targeting strategies begin with introduction of the vector into the fly genome. Different approaches can be used to get these "donor" flies, depending on the choice of vector.

3.3.1. P-Element-Based Transformation

Conventional *P*-element-mediated transgenesis (45) is used for vectors that do not carry *att*B sites (Fig. 2; see Note 11). Sequences between the P5′ and P3′ ends will be randomly incorporated into the fly genome. The insertions should be mapped to determine on which chromosome they are located before the crosses for the targeting event. This takes two generations, and plus time to establish a stock (4–6 weeks).

1. Cross G_0 flies (from injected embryos) with "double-balancer" flies carrying markers on chromosomes *2* and *3* in a

white mutant background. Integration of the targeting vector can be followed by the presence of the eye color conferred by the *mini-white* marker.

2. Backcross *white*+-balanced heterozygous progeny to double-balancer flies to map the insertion of the donor sequence by conventional genetic marker segregation.
3. G_2 generation: Establish a balanced stock for each insertion. Select the insertions that are not on the same chromosome as the gene to be targeted for subsequent homologous recombination.

3.3.2. ϕC31 Integrase-Mediated Transformation

ϕC31 is a phage integrase that catalyzes recombination between the nonidentical recognition sites, *att*P and *att*B sites. The recombination event produces two new sequences, *att*L and *att*R, which are not recognized by the integrase (46, 47). Hence, *ϕC31*-mediated recombination can generate stable integrant at a chromosomally resident *att*P site that will not be further excised or exchanged (48). *ϕC31* technology has proven to be highly efficient in *Drosophila* for targeting a plasmid carrying *att*B to a genomic *att*P landing site (41). By defining the chromosomal integration site, subsequent mapping steps are eliminated. Several labs have generated landing strains carrying *att*P site in the genome on all chromosomes (39–41, 44). pW25-*att*B vectors should be injected into a landing-site strain that carries *att*P site on the desired chromosome and also germ line-expressed integrase (e.g., *vasa-integrase*). G_0 flies from injected embryos should be crossed to appropriate balancer strains to establish stocks (see Note 12).

3.4. Gene Targeting

The crossing scheme use to generate the targeting event is shown in Fig. 3. The donor transgene is marked genetically by expression of the *white*+ marker. Transfer of the *white*+ marker from its starting location to the new location is used to monitor the progress of the targeting event. In some cases, the two sites can be distinguished by the intensity of the *white*+ eye color (this can be affected by integration site). However, this is not always the case. For this reason, it is convenient to start with a donor that is on a chromosome different from the locus to be targeted. This is not an absolute requirement but the genetics is simpler if done in this way (see Note 13). The example shown in Fig. 3 is for a locus to be targeted on chromosome *3*.

1. The targeting event is initiated by mobilizing the donor construct from its integration site. This is a two-step process involving use of FLP recombinase and the FRT sites in the plasmid to excise the construct as a circular DNA molecule. This is then linearized by cleavage at the I-*Sce*I endonuclease site in the vector. In practice, these two steps are done

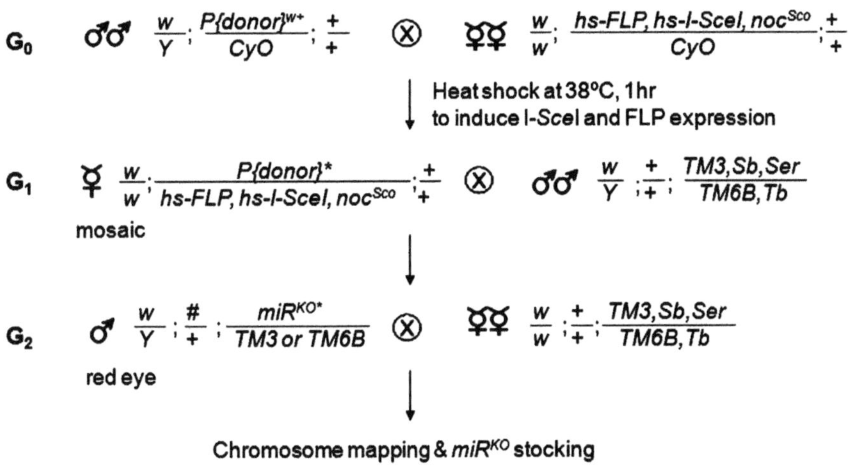

G_0 ♂♂ $\frac{w}{Y}; \frac{P\{donor\}^{w+}}{CyO}; \frac{+}{+}$ ⊗ ♀♀ $\frac{w}{w}; \frac{hs\text{-}FLP, hs\text{-}I\text{-}SceI, noc^{Sco}}{CyO}; \frac{+}{+}$

↓ Heat shock at 38°C, 1 hr to induce I-*Sce*I and FLP expression

G_1 ♀ $\frac{w}{w}; \frac{P\{donor\}^{*}}{hs\text{-}FLP, hs\text{-}I\text{-}SceI, noc^{Sco}}; \frac{+}{+}$ ⊗ ♂♂ $\frac{w}{Y}; \frac{+}{+}; \frac{TM3,Sb,Ser}{TM6B,Tb}$

mosaic

↓

G_2 ♂ $\frac{w}{Y}; \frac{\#}{+}; \frac{miR^{KO*}}{TM3\ or\ TM6B}$ ⊗ ♀♀ $\frac{w}{w}; \frac{+}{+}; \frac{TM3,Sb,Ser}{TM6B,Tb}$

red eye

↓

Chromosome mapping & *miR*^{KO} stocking

Fig. 3. The genetic crosses in ends-out gene targeting. Here is an example of targeting a miRNA on chromosome *3*, while a donor line ($P\{donor\}^{w+}$) on chromosome *2* is chosen. $P\{donor\}^{*}$ only transiently exists as a linearized extrachromosomal donor DNA fragment in the cell. "# chromosome" is inherited from the female. It could be the excised or un-excised donor, or *hs-FLP*, *hs-I-SceI*, or the recombinant in between the two chromosomes. miR^{KO*} is the potential targeting event. Donor flies are crossed to *hs-FLP*, *hs-I-SceI* flies. Set up ten crosses and flip them every day to a new tube for 5 consecutive days. Heat shock tubes on the third day at 38°C for 1 h. In the next generation, set up 400 crosses of individual female with mosaic eyes to balancer males. Screen for red eye males at G_2 generation. Red eye flies are further crossed to balancer flies and decide which chromosome it is on by marker segregation.

concurrently by induction of FLP and I-*Sce*I under heat shock control: Set up ten vials, each with 15–20 virgin females and 3–5 males. Transfer the adults to new vials every day for 1 week (or 5 days). Heat shock vial at 38°C for 1 h during first instar, 24–48 h after egg laying (on the second day after removing parent flies).

2. Excision of the donor DNA carrying the *white*$^{+}$ marker will occur in somatic as well as germ-line tissue. Flies in which this has occurred can be recognized by a mosaic pattern of white and pigmented ommatidia in the eyes (see Note 14).

3. The linearized targeting vector is naturally recombinogenic. In the event of recombination with the target locus in the germ line, the event can be passed on to the next generation. The efficiency of homologous recombination in female germ line is generally higher than in male germ line (22). To capture these events, female progeny in which donor excision has taken place (mosaic eyes) are crossed to males from a suitable balancer stock (in this example *white–* with two chromosome *3* balancers, Fig. 3). To ensure that independent targeting events are isolated we recommend setting up individual crosses with one or two females each, and selecting only one candidate per vial at the next generation. In practice, we routinely set up 400 such crosses to ensure a reasonable likelihood of success. Targeting efficiency is about 2% on average

for deletions of 100–200 bp. In the case of deletions >1 kb, the number of crosses might need to be greater (see Note 15).

4. The targeting event reintroduces the *white*+ eye color marker into the chromosome. Candidates can be selected as progeny with *white*+ eyes at this generation. If possible, we select male progeny, and only one event per vial, to ensure than all candidate lines are independent (see Note 16). This is important because not all *white*+ progeny will be the result of the desired event. Excision of the donor is not generally 100% efficient and a variable proportion of the progeny may carry the donor vector. These can be easily distinguished from the desired targeting event if the eye colors are different. Nonhomologous integration can also lead to insertion of the plasmid (or part thereof) at random locations. The first step to determine which of the progeny may be the desired mutant is to map the *white*+ eye color to the correct chromosome (chromosome *3* in the example in Fig. 3). Backcrossing individual *white*+*/Balancer* male progeny to third chromosome balancer females (*white*−) for one generation allows mapping to ensure the *white*+ transgene is on the correct chromosome and to establish a balanced stock.
5. Although a balanced mutant strain can be maintained from this stage, it is advisable to first clean the chromosome. The targeted allele is marked by the *white*+ transgene and so can easily be recovered after backcrossing to a *white*− strain. Several generations of backcrossing for free recombination and chromosome exchange are recommended. In the case of closely linked mutations, more active cleaning methods may be required, such as multiple recessive marker recombination or male recombination (21). To be certain that any mutant phenotype observed is due to the targeting event, there is no substitute for genetic rescue. Several of the vectors have been designed to facilitate rescue experiment (e.g., by RMCE). We consider confirmation by rescue to be an essential step in any targeting strategy.

3.5. Molecular Verification of Targeted Knockouts

Molecular verification of the targeting event is essential. In most cases, the desired event is the most common outcome, once the transgene has been mapped to the correct chromosome. 80% of targeted events have been validated as being molecularly correct in our experience. The 20% of failed cases include instances of nonhomologous integration of the transgene. In rare cases, more complex outcomes have been observed involving multiple copies of the transgene. Two levels of validation of transgene location are recommended.

3.5.1. Verify that the White Marker Is at the Target Locus

The presence of the *mini-white* gene can be verified by PCR on genomic DNA from heterozygous targeted flies. Note that the flanking region primer (pflank) should be located outside the

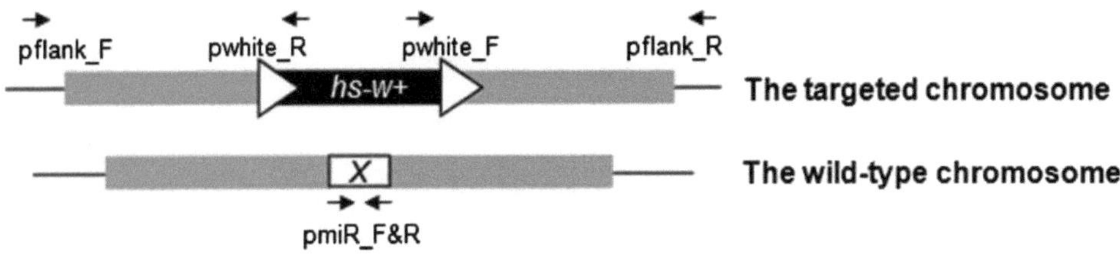

Fig. 4. An illustration for primer design in molecular verification of targeted mutants. To verify the presence of the *mini-white* marker, a forward primer (pflank_F) in the flanking region located outside of homology arm is used to pair with a reverse primer (pwhite_R) in the *mini-white* sequence for amplification of left homology arm. Similarly for the right arm, a forward primer (pwhite_F) in the *mini-white* sequence and a reverse primer (pflank_R) in the flanking region located outside of homology region are used for amplification. To verify the loss of miRNA in targeted mutants, a pair of primers (pmiR_F and R) in the deleted region is used for amplification. The amplicon is missing in the homozygous mutant flies but not in the nonmutated flies.

homology arm (see Fig. 4). The second primer (pwhite) comes from the *mini-white* sequence. PCR products of 4 kb or greater are needed, so good quality genomic DNA is essential (see Note 17). Amplification at both ends of the *mini-white* gene should be confirmed.

3.5.2. Verify Loss of the miRNA

In addition to confirming that the transgene is inserted at the correct position, it is advisable to confirm molecularly the absence of the sequences that were to be targeted for deletion. This can be done conveniently by PCR from genomic DNA using primers specific to the deleted region, assuming that the mutant allele is viable so that DNA can be obtained from homozygous mutants (Fig. 4). If this material is difficult to obtain, single-fly PCR may be needed (see Note 18). Alternatively, the presence or absence of the miRNA can be determined by miRNA quantitative PCR. We have had good success with the TaqMan MicroRNA Assay using a quantitative PCR machine (Applied Biosystems), but other commercial miRNA quantification systems are also useful.

If homozygous mutant material cannot be obtained, it is possible to carry out quantitative PCR on heterozygous DNA or RNA samples. A twofold difference in the amount of DNA or mature miRNA transcript is expected when comparing the heterozygous mutant sample to a control. We have used this approach with success, but additional controls are needed, such as an independent heterozygous deletion strain uncovering the locus. This approach to validation should be considered as a last resort. Of course, conventional methods such as Southern Blots can also be used to detect the genomic deletion as another means to verify heterozygosity.

3.6. Generating "Clean" Mutations by Marker Removal

The *mini-white* gene used as a genetic marker is mostly comprised of genomic sequences and so has its own exons and introns (49). Splice acceptor sites in *mini-white* can interfere with splicing

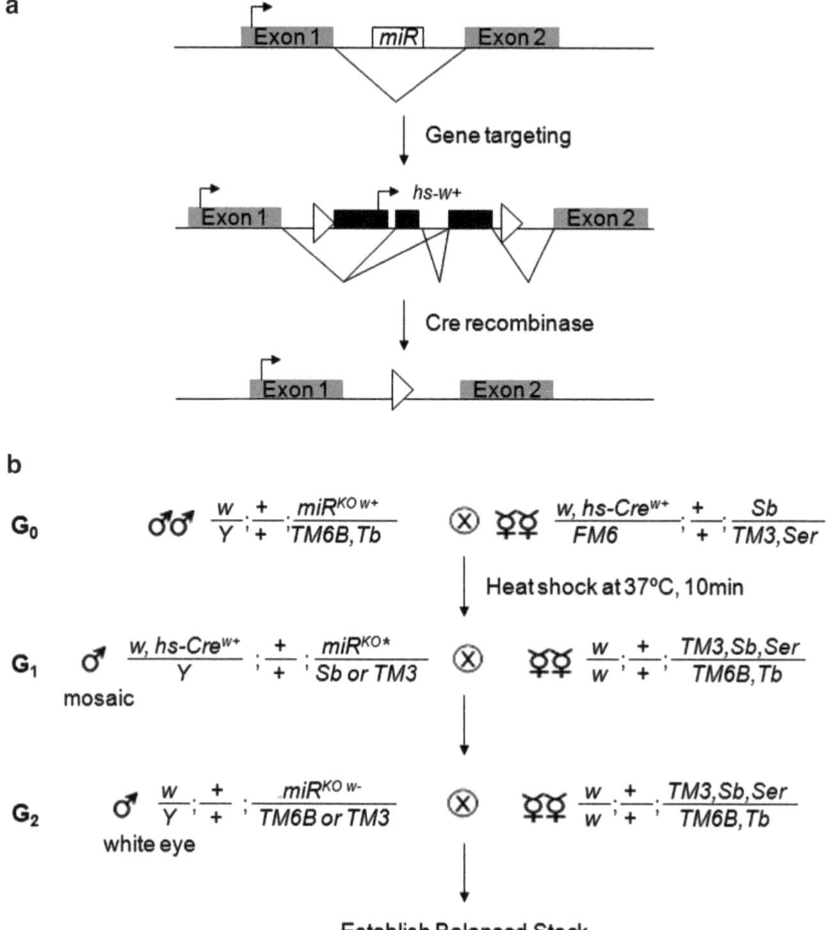

Fig. 5. Removing *mini-white* marker in a case of intronic miRNA. (**a**) A schematic demonstration of gene targeting and marker removal for an intronic miRNA. A miRNA located in the intron of its host gene is replaced by *mini-white* flanked by *lox*P sites during gene-targeting process. The genomic sequence of *white* gene driven by *hsp70* promoter (*hs-w*$^+$) is represented here by three exons, denoted by *black boxes* (as there are six exons in reality). The abnormal splicing isoform of the host gene incorporates the exons from *white* gene and disrupts its normal function. With the expression of Cre recombinase which can excise the sequence flanked by *lox*P sites, only one *lox*P site is left behind and would not disrupt splicing of the host gene anymore. (**b**) The genetic crosses for *mini-white* excision in miRNA knockout flies by Cre/*loxP* system. Here is an example of miRNA knockout mutant (*miR*$^{KO\ w+}$) on chromosome *3*, using an *X* chromosome source of *hs-Cre*. Knockout flies are crossed to flies carrying *hs-Cre*$^{w+}$, which is also marked by red eye. In the next generation, select males with mosaic eyes on an orange background and cross them to balancer flies individually. In G_2 generation, white eye males are selected for crosses and stocking.

of nearby genes if the targeting event places the *mini-white* reporter in the correct orientation in an intron (see Fig. 5a). The "gene trap" nature of the targeting vector should be considered when making miRNA mutants if the miRNA is located in the intron of a protein-coding gene (as is the case for 57 out of 152 miRNAs). In such cases, the miRNA targeting event will also mutate the host gene. The *mini-white* reporter is flanked by *loxP* sites, which can be used to remove the marker using Cre recombinase (24, 43). Following recombination, only one *loxP* site will be left in the intron. This would not be expected to affect expression of the "host" gene. Gene traps can be cured efficiently with the following simple protocol (Fig. 5b).

1. Cross flies carrying the targeted allele with Cre recombinase flies. Expression of Cre recombinase is under control of the heat shock promoter. First instar larvae from this cross are given heat shock treatment for 10 min at 37°C to induce Cre expression (i.e., 24–48 h after egg laying; see Note 19).
2. Excision of *mini-white* can be identified by loss of eye pigment. This is often mosaic in the somatic tissue and in the germ line. Collect mosaic-eyed males and set up multiple single crosses with balancer flies. The desired progeny from this cross will have white eyes. These can be used to establish a balanced stock by backcrossing to an appropriately marked balancer strain, as illustrated in Fig. 5b.
3. It is advisable to confirm the excision of the complete cassette and loss of the miRNA sequences by PCR on the stock, as discussed above.

3.7. Genetic Rescue

Mobilizing transposons and use of recombinases can be mutagenic. It is not uncommon to induce lesions in addition to the desired targeting event. It is therefore advisable to clean the targeted genome to remove unlinked mutations, and to clean the targeted chromosome by meiotic recombination to remove linked mutations. We strongly recommend that mutant phenotype(s) be confirmed by genetic rescue. Suppression of the mutant phenotype upon re-expression of the miRNA provides compelling evidence that the original defect is caused by deletion of the miRNA. Various approaches are possible, and the best choice may depend on the targeting vector used. The rescue construct should be able to restore miRNA expression at a level comparable to the endogenous level, ideally with a normal spatio-temporal profile. A genomic rescue construct covering the miRNA locus and its endogenous cis-regulatory sequences region is ideal. In cases where *Gal4* series vectors are used to express *Gal4* under control of the endogenous miRNA regulatory elements, a *UAS-miRNA* constructs can be used for rescue. In some cases, "leaky" *Gal4* -independent expression of a *UAS-miRNA* transgene may be sufficient for rescue. Levels of "leaky" expression can be monitored by quantitative miRNA PCR. A third approach to rescue is based on use of RMCE to introduce the miRNA back into the mutated locus.

3.8. Recombinase-Mediated Cassette Exchange

Recombinase-mediated cassette exchange (RMCE) with the pW25-RMCE vector provides a versatile tool for sequential manipulation of a targeted locus. The principle is illustrated in Fig. 6.

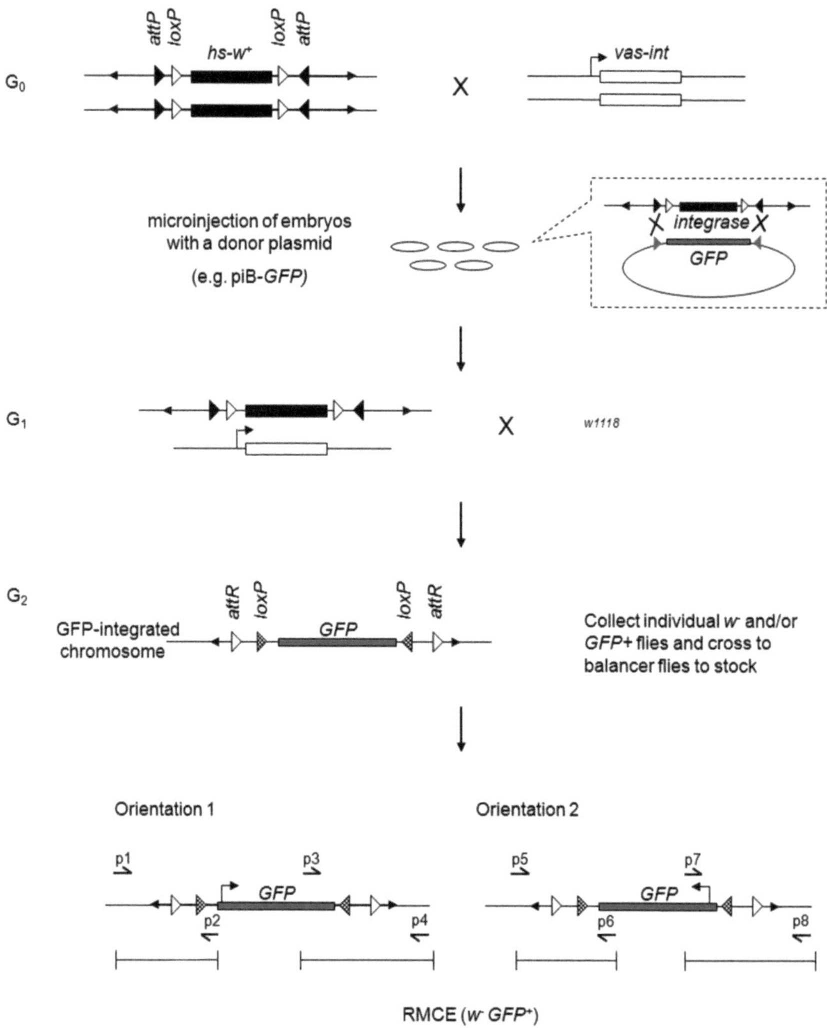

Fig. 6. Schematic illustration of the application of RMCE in mutant genome manipulation. Flies carrying the targeted locus, marked by *mini-white*, are crossed to transgenic flies expressing *ϕC31* integrase. Embryos obtained from this cross are injected with a donor plasmid expressing genetic sequence of interest flanked by inverted *att*B sites, such as piB-GFP donor plasmid, which expresses *GFP*. Double cross-over between the two *att*P and *att*B sites would lead to a clean exchange of *mini-white* by *GFP*. Emerging adults are mated individually with w^{1118} partners and putative RMCE events were identified by the loss of *mini-white* expression in the eyes. Molecular genotyping by PCR is used to determine the orientation of *GFP* insertion.

Sequences between the inverted *att*P sites can be efficiently exchanged with a plasmid-borne sequence flanked by *att*B sites in the presence of bacteriophage *ϕC31* integrase. This can be used to replace the *mini-white* gene trap with the miRNA hairpin, for genetic rescue experiments. RMCE can also be used to "cure" a gene trap by exchanging an intron-less *GFP* reporter for *mini-white*. In our experience, RMCE can occur with efficiencies of up to ~50% of surviving injected flies producing an exchange event (37).

1. Flies carrying the locus to be re-targeted by RMCE are crossed to flies expressing *ϕC31* integrase in the germ line. The resulting embryos contain integrase and can be injected with the donor *att*B plasmid. The example in Fig. 6 shows use of the piB-GFP plasmid (38).
2. Adult flies emerging from these embryos are crossed individually to a *white* mutant strain. In this example, use of w^{1118} allows identification of flies from which the *mini-white* marker was successfully excised. Screen for putative RMCE candidates by the loss of eye pigment and/or gain of *GFP* expression in the case of injection with piB-GFP plasmid (see Note 20). These should possess the targeted allele, containing the sequence, such as *GFP*, introduced from the donor plasmid in place of *mini-white*. Collect individual flies and cross them to appropriate flies to establish balanced stocks (G_2 generation).
3. Verify RMCE candidates by PCR and the direction of the newly introduced sequence using molecular analyses such as PCR. The orientation of cassette insertion is random, so the PCR strategy should be designed to allow detection of both possible outcomes, as illustrated in Fig. 6.

4. Notes

1. In the case of miRtrons, in which the mature miRNA is derived by a different mechanism, involving splicing of a normal protein-coding transcript (50, 51), deleting the entire hairpin might affect the splicing sites of the host transcript. The mature sequence of the miRNA can be targeted instead.
2. There are many sequence differences between the reference genome and the commonly used lab strains – so it is a good idea to sequence verify your strain at the targeted locus.
3. The two translation stop cassettes are designed to ensure that insertional disruption alleles are still mutant after marker removal in a protein-coding sequence.
4. The *Asc*I site in the 3′ MCS has been replaced by *Sbf*I, so downstream homology arm cloning strategies may differ from the other vectors.
5. Additional *Pac*I and *Fse*I sites were introduced to the 5′ and 3′ multiple cloning sites, to facilitate directional cloning of the homology arms flanking the targeted locus.
6. Targeting efficiency also depends greatly on the amount of DNA to be deleted. Homology arms on the order of 3.5–4 kb are usually sufficient to allow small deletions of the sort needed to remove a miRNA (~100 bp) with good efficiency.

The shortest homology arm we have tried successfully is 2.5 kb, but we have not made an extensive effort to push the limits on this parameter. It may be that the amount of homology needed varies with the locus to be targeted.

7. Pay attention to which strand the miRNA is expressed from when designing the vector. The orientation will not matter for simple deletions, using pW25, but may be important for strategies involving *Gal4* or *GFP* knock-in (e.g., using pW25-Gal4 or pW25-RMCE).
8. *Not*I and *Asc*I restriction enzymes have GC-rich eight-base target sequences. Sites are found comparatively rarely in the fly genome, so we have found it convenient to incorporate these sites into the PCR primers used to amplify the homology arms. Be sure to check for the absence of these sites in your homology arms. If your target gene has one of these sites nearby, *PspOM*I, *Sph*I, or *Kpn*I are alternative choices for cloning into the 5′ MCS. *BsiW*I is an alternative for downstream MCS cloning in pW25. Restriction-enzyme-free cloning methods, such as SLIC (sequence- and ligation-independent cloning) provides another means to circumvent use of enzyme sites (52). Recombineering strategies may also be useful (44).
9. In our hands, the error rate from Phusion polymerase is slightly lower than the other two *Taq* DNA polymerases. We recommend using fewer than 30 cycles to minimize the error rate for all enzymes.
10. It is the sequence of your lab strain that matters for efficient recombination with the homology arms.
11. Targeting vectors are typically about 17 kb in length (~10 kb of vector sequences plus 2 homology arms of ~3–4 kb). Their size will affect the efficiency of *P*-element-mediated transformation but in practice typically not to a degree that compromises the procedure. The integrase-mediated transgenesis vectors of the pW25-*att*B series (Fig. 2d–f) were designed to allow efficient introduction on the targeting vector to the desired chromosome (37).
12. We have successfully used the following landing strains on chromosome *2* and *3*: *att*P2, *att*P16, *ZH-22A*, *ZH-86Fb*. *ZH-86Fb* has proven to be the most efficient. Viability after injection has been low for *ZH-22A* compared to the other three landing-site strains.
13. The FLP/FRT system can also be used to distinguish the donor from the targeted event (23, 34), as follows: The donor is flanked by FRT sites that are used to excise the element from the chromosome to generate a circular DNA molecule as an intermediary in the targeting process. The *white*+ marker at a properly targeted locus cannot be removed by FLP, and

so these flies will not produce mosaic eyes, when exposed to FLP. Note that FRT sites can be damaged, leading to failed excision. FRT-excision mosaicism will fail to distinguish damaged donors from correctly targeted events, so this method is not as reliable as chromosome mapping.

14. It is possible for all ommatidia to be phenotypically *white*⁻ if excision has been efficient. Efficiency will vary with the donor.

15. Not all donors are equal. Our experience has been that 17% of donors failed to produce targeting events. In most of these cases, the problem could be traced to the original donor line, and crosses with an independently isolated donor insertion line were successful. We recommend using multiple independent donor inserts, if available (e.g., 200 crosses each with two donors, for a total of 400 crosses). Donor-site differences do not pertain when using targeted integration site donors, but an individual integrant can sometimes be damaged, so it is always advisable to start with multiple donors.

16. In the case of targeting miRNAs on *X* chromosome, bear in mind that some knockout events can only be recovered in females (which are heterozygous for the targeted locus). If the mutation is lethal, mutant males will not be recovered. We advise using both male and female progeny when making mutants on the *X* chromosome.

17. Since heterozygous flies are a suitable source of DNA for this purpose, it is advisable to start with plenty of material. The Qiagen Tissue Extraction Kit has proven to be suitable, but alternative sources of reagents also work well.

18. Place a single fly into a microfuge tube and leave on ice. Add 10 mg/ml Proteinase K into SB freshly. Take 50 μl of SB and grind the fly using the yellow tip for 20 s without pipetting out the solution. Remove the solution and incubate at 37°C for 30 min. Incubate at 95°C for 2 min to inactivate Proteinase K. Samples can be stored at 4°C for up to 1 month. PCR is performed with primers for the miRNA stem loop inside the deleted region (Fig. 4). Product amplification would not be expected for homozygous mutants but is expected in nonmutated samples. DNA from male flies is preferred, to avoid ambiguity due to stored sperm in nonvirgin females.

19. High-level expression of Cre recombinase causes lethality. Low-level expression under *hsp70* promoter at 25°C can be sufficient to induce recombination. G_1 progeny may have mosaic eyes and produce white-eyed progeny in the next generation. If the *hs-Cre* transgene chosen is marked by *mini-white* (as illustrated in Fig. 5b), the mosaicism will be on an orange background. There are other sources of *hs-Cre*, some

of which are marked by *yellow*[+] (available in Bloomington Stock Center). Then, the mosaicism will be on white background.

20. The *ϕC31* integrase transgene is tagged with 3P-GFP. Therefore, flies expressing the *ϕC31* integrase also express *GFP* in the eyes. This expression is very strong, and easily scored under a binocular *GFP* microscope. Effort should be made to outcross this integrase-carrying allele to avoid confusion with expression by the knocked-in *GFP* reporter.

Acknowledgments

We thank Rubing Liu, Hai Hwee Tay, Kah Junn Tan, and Yoke Ping Gum for technical support. Susan from Genetic Services Inc. provided injection services ably and with patience when needed. Natascha Bushati and Boris Bryk helped by sharing their experiences in generating miRNA knockouts. We thank Dr. Pernille Rorth for providing *hs-Cre* strains. This work has been supported by EU-FP6 grant "Sirocco" LSHG-CT-2006-037900, Singapore National Research Foundation under CRP Award No. NRF-CRP3-2008-03, and Temasek Life Sciences Laboratory. Ruifen Weng is a recipient of a Singapore Millennium Foundation Scholarship.

References

1. Bartel, D. P. (2009) MicroRNAs: target recognition and regulatory functions. *Cell* **136**, 215–33.
2. Flynt, A. S., and Lai, E. C. (2008) Biological principles of microRNA-mediated regulation: shared themes amid diversity. *Nat Rev Genet* **9**, 831–42.
3. Bushati, N., and Cohen, S. M. (2007) microRNA functions. *Annu Rev Cell Dev Biol* **23**, 175–205.
4. Lee, R. C., Feinbaum, R. L., and Ambros, V. (1993) The C. elegans heterochronic gene lin-4 encodes small RNAs with antisense complementarity to lin-14. *Cell* **75**, 843–54.
5. Wightman, B., Ha, I., and Ruvkun, G. (1993) Posttranscriptional regulation of the heterochronic gene lin-14 by lin-4 mediates temporal pattern formation in C. elegans. *Cell* **75**, 855–62.
6. Rorth, P., Szabo, K., Bailey, A., Laverty, T., Rehm, J., Rubin, G. M., Weigmann, K., Milan, M., Benes, V., Ansorge, W., and Cohen, S. M. (1998) Systematic gain-of-function genetics in Drosophila. *Development* **125**, 1049–57.
7. Brennecke, J., Hipfner, D. R., Stark, A., Russell, R. B., and Cohen, S. M. (2003) bantam encodes a developmentally regulated microRNA that controls cell proliferation and regulates the proapoptotic gene hid in Drosophila. *Cell* **113**, 25–36.
8. Xu, P., Vernooy, S. Y., Guo, M., and Hay, B. A. (2003) The Drosophila microRNA Mir-14 suppresses cell death and is required for normal fat metabolism. *Curr Biol* **13**, 790–5.
9. Li, X., and Carthew, R. W. (2005) A microRNA mediates EGF receptor signaling and promotes photoreceptor differentiation in the Drosophila eye. *Cell* **123**, 1267–77.
10. Karres, J. S., Hilgers, V., Carrera, I., Treisman, J., and Cohen, S. M. (2007) The conserved microRNA miR-8 tunes atrophin levels to prevent neurodegeneration in Drosophila. *Cell* **131**, 136–45.

11. Ketting, R. F., Fischer, S. E., Bernstein, E., Sijen, T., Hannon, G. J., and Plasterk, R. H. (2001) Dicer functions in RNA interference and in synthesis of small RNA involved in developmental timing in C. elegans. *Genes Dev* **15**, 2654–9.
12. Knight, S. W., and Bass, B. L. (2001) A role for the RNase III enzyme DCR-1 in RNA interference and germ line development in Caenorhabditis elegans. *Science* **293**, 2269–71.
13. Grishok, A., Pasquinelli, A. E., Conte, D., Li, N., Parrish, S., Ha, I., Baillie, D. L., Fire, A., Ruvkun, G., and Mello, C. C. (2001) Genes and mechanisms related to RNA interference regulate expression of the small temporal RNAs that control C. elegans developmental timing. *Cell* **106**, 23–34.
14. Hatfield, S. D., Shcherbata, H. R., Fischer, K. A., Nakahara, K., Carthew, R. W., and Ruohola-Baker, H. (2005) Stem cell division is regulated by the microRNA pathway. *Nature* **435**, 974–8.
15. Jin, Z., and Xie, T. (2007) Dcr-1 maintains Drosophila ovarian stem cells. *Curr Biol* **17**, 539–44.
16. Wienholds, E., Koudijs, M. J., van Eeden, F. J., Cuppen, E., and Plasterk, R. H. (2003) The microRNA-producing enzyme Dicerl is essential for zebrafish development. *Nat Genet* **35**, 217–8.
17. Murchison, E. P., Partridge, J. F., Tam, O. H., Cheloufi, S., and Hannon, G. J. (2005) Characterization of Dicer-deficient murine embryonic stem cells. *Proc Natl Acad Sci USA* **102**, 12135–40.
18. Kanellopoulou, C., Muljo, S. A., Kung, A. L., Ganesan, S., Drapkin, R., Jenuwein, T., Livingston, D. M., and Rajewsky, K. (2005) Dicer-deficient mouse embryonic stem cells are defective in differentiation and centromeric silencing. *Genes Dev* **19**, 489–501.
19. Cayirlioglu, P., Kadow, I. G., Zhan, X., Okamura, K., Suh, G. S., Gunning, D., Lai, E. C., and Zipursky, S. L. (2008) Hybrid neurons in a microRNA mutant are putative evolutionary intermediates in insect CO2 sensory systems. *Science* **319**, 1256–60.
20. Robertson, H. M., Preston, C. R., Phillis, R. W., Johnson-Schlitz, D. M., Benz, W. K., and Engels, W. R. (1988) A stable genomic source of P element transposase in Drosophila melanogaster. *Genetics* **118**, 461–70.
21. Preston, C. R., Sved, J. A., and Engels, W. R. (1996) Flanking duplications and deletions associated with P-induced male recombination in Drosophila. *Genetics* **144**, 1623–38.
22. Rong, Y. S., and Golic, K. G. (2000) Gene targeting by homologous recombination in Drosophila. *Science* **288**, 2013–8.
23. Rong, Y. S., and Golic, K. G. (2001) A targeted gene knockout in Drosophila. *Genetics* **157**, 1307–12.
24. Rong, Y. S., Titen, S. W., Xie, H. B., Golic, M. M., Bastiani, M., Bandyopadhyay, P., Olivera, B. M., Brodsky, M., Rubin, G. M., and Golic, K. G. (2002) Targeted mutagenesis by homologous recombination in D. melanogaster. *Genes Dev* **16**, 1568–81.
25. Gong, W. J., and Golic, K. G. (2003) Ends-out, or replacement, gene targeting in Drosophila. *Proc Natl Acad Sci USA* **100**, 2556–61.
26. Gong, W. J., and Golic, K. G. (2004) Genomic deletions of the Drosophila melanogaster Hsp70 genes. *Genetics* **168**, 1467–76.
27. Xie, H. B., and Golic, K. G. (2004) Gene deletions by ends-in targeting in Drosophila melanogaster. *Genetics* **168**, 1477–89.
28. Sokol, N. S., and Ambros, V. (2005) Mesodermally expressed Drosophila microRNA-1 is regulated by Twist and is required in muscles during larval growth. *Genes Dev* **19**, 2343–54.
29. Teleman, A. A., Maitra, S., and Cohen, S. M. (2006) Drosophila lacking microRNA miR-278 are defective in energy homeostasis. *Genes Dev* **20**, 417–22.
30. Li, Y., Wang, F., Lee, J. A., and Gao, F. B. (2006) MicroRNA-9a ensures the precise specification of sensory organ precursors in Drosophila. *Genes Dev* **20**, 2793–805.
31. Bushati, N., Stark, A., Brennecke, J., and Cohen, S. M. (2008) Temporal reciprocity of miRNAs and their targets during the maternal-to-zygotic transition in Drosophila. *Curr Biol* **18**, 501–6.
32. Friggi-Grelin, F., Lavenant-Staccini, L., and Therond, P. (2008) Control of antagonistic components of the hedgehog signaling pathway by microRNAs in Drosophila. *Genetics* **179**, 429–39.
33. Sokol, N. S., Xu, P., Jan, Y. N., and Ambros, V. (2008) Drosophila let-7 microRNA is required for remodeling of the neuromusculature during metamorphosis. *Genes Dev* **22**, 1591–6.
34. Maggert, K. A., Gong, W. J., and Golic, K. G. (2008) Methods for homologous recombination in Drosophila. *Methods Mol Biol* **420**, 155–74.
35. Gao, G., McMahon, C., Chen, J., and Rong, Y. S. (2008) A powerful method combining homologous recombination and site-specific

recombination for targeted mutagenesis in Drosophila. *Proc Natl Acad Sci USA* **105**, 13999–4004.

36. Choi, C. M., Vilain, S., Langen, M., Van Kelst, S., De Geest, N., Yan, J., Verstreken, P., and Hassan, B. A. (2009) Conditional mutagenesis in Drosophila. *Science* **324**, 54.
37. Weng, R., Chen, Y. W., Bushati, N., Cliffe, A., and Cohen, S. M. (2009) Recombinase-mediated cassette exchange provides a versatile platform for gene targeting: knockout of miR-31b. *Genetics* **183**, 399–402.
38. Bateman, J. R., Lee, A. M., and Wu, C. T. (2006) Site-specific transformation of Drosophila via phiC31 integrase-mediated cassette exchange. *Genetics* **173**, 769–77.
39. Bischof, J., Maeda, R. K., Hediger, M., Karch, F., and Basler, K. (2007) An optimized transgenesis system for Drosophila using germ-line-specific phiC31 integrases. *Proc Natl Acad Sci USA* **104**, 3312–7.
40. Markstein, M., Pitsouli, C., Villalta, C., Celniker, S. E., and Perrimon, N. (2008) Exploiting position effects and the gypsy retrovirus insulator to engineer precisely expressed transgenes. *Nat Genet* **40**, 476–83.
41. Groth, A. C., Fish, M., Nusse, R., and Calos, M. P. (2004) Construction of transgenic Drosophila by using the site-specific integrase from phage phiC31. *Genetics* **166**, 1775–82.
42. Gloor, G. B., Preston, C. R., Johnson-Schlitz, D. M., Nassif, N. A., Phillis, R. W., Benz, W. K., Robertson, H. M., and Engels, W. R. (1993) Type I repressors of P element mobility. *Genetics* **135**, 81–95.
43. Siegal, M. L., and Hartl, D. L. (1996) Transgene Coplacement and high efficiency site-specific recombination with the Cre/loxP system in Drosophila. *Genetics* **144**, 715–26.
44. Venken, K. J., He, Y., Hoskins, R. A., and Bellen, H. J. (2006) P[acman]: a BAC transgenic platform for targeted insertion of large DNA fragments in D. melanogaster. *Science* **314**, 1747–51.
45. Spradling, A. C. (1986) P element-mediated transformation *in* "Drosophila: a practical approach" (Roberts, D. B., Ed.), pp. 175–97, IRL Press Limited, Oxford, England.
46. Kuhstoss, S., and Rao, R. N. (1991) Analysis of the integration function of the streptomycete bacteriophage phi C31. *J Mol Biol* **222**, 897–908.
47. Rausch, H., and Lehmann, M. (1991) Structural analysis of the actinophage phi C31 attachment site. *Nucleic Acids Res* **19**, 5187–9.
48. Thorpe, H. M., Wilson, S. E., and Smith, M. C. (2000) Control of directionality in the site-specific recombination system of the Streptomyces phage phiC31. *Mol Microbiol* **38**, 232–41.
49. Klemenz, R., Weber, U., and Gehring, W. J. (1987) The white gene as a marker in a new P-element vector for gene transfer in Drosophila. *Nucleic Acids Res* **15**, 3947–59.
50. Ruby, J. G., Jan, C. H., and Bartel, D. P. (2007) Intronic microRNA precursors that bypass Drosha processing. *Nature* **448**, 83–6.
51. Okamura, K., Hagen, J. W., Duan, H., Tyler, D. M., and Lai, E. C. (2007) The mirtron pathway generates microRNA-class regulatory RNAs in Drosophila. *Cell* **130**, 89–100.
52. Li, M. Z., and Elledge, S. J. (2007) Harnessing homologous recombination in vitro to generate recombinant DNA via SLIC. *Nat Methods* **4**, 251–6.

Chapter 9

Engineering Elements for Gene Silencing: The Artificial MicroRNAs Technology

Pablo Andrés Manavella and Ignacio Rubio-Somoza

Abstract

Small RNA (sRNA)-mediated gene silencing constitutes a powerful tool for the molecular characterization of a given gene. The RNAi technology has been largely used for this purpose. This approach is based on the cloning of an inverted repeated fragment of the gene to be silenced. Even when this approach produces a strong repression of the target gene it also involves the production of multiple small RNAs species that can easily lead to off targeting. Taking advantage of the latest insights into the new post-biogenesis layer of regulation in microRNA (miRNA) activity, it is possible to overcome the above-mentioned limitation. Artificial microRNAs (amiRNAs) are 21mer small RNAs, which can be genetically engineered and they function to specifically silence single or multiple genes of interest. Since generally just one miRNA molecule is generated from each precursor, the specificity of this technology is much higher than longer inverted repeats. Application of this technology results in highly specific mRNA downregulation by computationally designed sequences programmed to target one or a set of custom-selected transcripts.

Key words: MicroRNA, Small RNA attenuation, Specific gene downregulation, Gene silencing

1. Introduction

In plants, directed gene silencing has often been achieved by the use of small interfering RNAs (siRNA) to trigger RNAi. To trigger gene silencing, siRNA fragments of 200–800 bp of a gene of interest are cloned in both orientations separated by a loop-forming intron sequence. Although this approach has been successfully used to silence numerous genes, it has a major disadvantage because many siRNAs are formed from the original precursor. This makes it difficult to predict the cross silencing of non-intended targets and therefore challenging to ensure the silencing of individual members of a multigene family. However, artificial microRNAs are produced from miRNA precursors that usually generate only one single stable small RNA molecule, and the complete spectrum of

Tamas Dalmay (ed.), *MicroRNAs in Development: Methods and Protocols*, Methods in Molecular Biology, vol. 732,
DOI 10.1007/978-1-61779-083-6_9,

amiRNA targets is easily predictable (1). After sRNA production, the silencing of the target genes by siRNA and amiRNAs is very similar and occurs through cleavage-mediated transcript degradation, although translational inhibition of the target mRNA can also be observed (2). The use of amiRNA to silence specific genes has been described for numerous plant species including *Arabidopsis thaliana*, *Oryza sativa*, and *Physcomitrella patens* and also for the unicellular alga *Chlamydomonas reinhardtii* (3–7).

2. Materials

1. The template plasmids, containing the miR319a precursor (pRS300) or miR528 precursor (pNW55), are available under requested from Prof. Detlef Weigel (weigel@tuebingen.mpg.de).
2. Primer A: 5′-CTGCAAGGCGATTAAGTTGGGTAAC-3′ and primer B: 5′-GCGGATAACAATTTCACACAGGAAACAG-3′ are required for amiRNA cloning.
3. Any PCR product purification kit is suitable for the cloning process. In our case, we use and recommend the Wizard PCR purification kit (Promega).
4. Cloning vectors such as pGEM-T easy (promega) or pCR-TOPO (invitrogen) are required for the amiRNA cloning. Alternatively, gateway cloning vectors such as pCR8-GW-Topo (invitrogen) can be used, if the subcloning approach includes gateway recombination.
5. Invitrogen GeneRacer Kit (L1502-01) is required to determine the amiRNA cleavage product of any endogenous mRNA.
6. PCR reagents and equipment.
7. Appropriate binary vectors are required for subcloning of the amiRNA construct and its transformation into plants.

3. Methods

3.1. Design of the amiRNA

Artificial microRNAs (amiRNAs) are single-stranded 21mer RNAs with complementarity to one or several endogenous genes. An endogenous miRNA precursor is required as a backbone for assembling the amiRNA to ensure its correct processing. For *Arabidopsis*, we successfully used the MIR319a precursor to engineer amiRNAs by replacing the original miR319a with the artificial sequence and replacing the miR319a* with a sequence that pairs to the amiRNA. The miR319a locus has been described in detail and the use of its precursor as a backbone results in a high production of amiRNA molecules, leading to a strong silencing of the target gene (3). Additional miRNA precursors, such as the miRNA159 precursor, have been successfully used to engineer amiRNAs (8, 9). *Arabidopsis* precursors have been successfully used in tomato and tobacco (10), but we recommend testing the endogenous miRNA precursors from the species of interest for amiRNA engineering. Preference should be given to miRNA precursors for which expression data is already available. To obtain highly effective amiRNAs, one should design the constructs to resemble natural miRNAs as much as possible. Therefore several criteria should be followed:

1. The identity of the nucleotide on the 5′ end of the miRNA is responsible for loading it into the appropriate AGO1 complex and leads to target mRNA silencing (11). Since most plant and animal miRNAs start with a U at the 5′ end, amiRNA should also include this base at this position.
2. The amiRNA should display 5′ instability relative to the amiRNA*. This instability is critical for the selection of which miRNA/miRNA* strand is loaded into the AGO complex. In the double-strand duplex, the strand with lower thermodynamic stability at its 5′ end is preferentially incorporated into RISC (12, 13).
3. There should be no mismatch between positions 2 and 12 of the amiRNA for all targets. Although a few mismatches can be tolerated for multiple targets, no mismatches in positions 10 and 11 are allowed under any circumstance. Perfect complementarity in positions 10 and 11 between miRNA and target are essential for target cleavage (3).
4. Even when there is no evidence for transitive formation of secondary siRNAs from the target gene after amiRNA cleavage, it is recommended to include one or two mismatches at the amiRNA 3′ end (positions 18–21) to reduce this possibility.
5. It is advisable that the amiRNA present a similar mismatch pattern for all intended targets to reduce variability between targets.
6. Absolute hybridization energy between −35 and −38 kcal/mol are recommended since these are the values observed for most endogenous miRNA targets. Do not consider any amiRNA which pair to intended targets with energies higher than −30 kcal/mol.

The criteria to be considered in the design of an amiRNA are complex and the use of a computational algorithm facilitates the process. With this purpose, the Detlef Weigel lab has developed an online tool set to compute optimal amiRNA sequences.

The Web MicroRNA Designer version 3 (WMD3) is available online (http://wmd3.weigelworld.org/) and allows users to generate all possible specific amiRNAs just by providing the target gene accession data or the sequence. Additionally, once the amiRNA sequence has been selected, the WMD3 tool provides the sequences of the primers needed to engineer the amiRNA into an endogenous microRNA precursor. The tool designs primers for the *A. thaliana* miR319a (ath-miR319a) and *O. sativa* miR528 (osa-miR528) precursors.

3.2. Cloning amiRNAs

1. The artificial microRNA designer WMD3 delivers four oligonucleotide sequences (I–IV), which are used to engineer your artificial microRNA into the endogenous miR319a/miR528 precursor by site-directed mutagenesis.
2. Two additional oligonucleotides based on the template plasmid sequence are needed (primer A: 5′-CTGCAAGGCGATT AAGTTGGGTAAC-3′ and primer B: 5′-GCGGATAACAATT TCACACAGGAAACAG-3′). They are located outside of the multiple cloning site of plasmid to generate PCR products easily cloned using the polycloning site of the original vector.
3. The amiRNA-containing precursor is generated by overlapping PCR as shown in Fig. 1. To build this precursor, follow these steps:
   1. Using the pRS300 as template, amplify three fragments by PCR using primers A + IV, primers II + III, and primers I + B.
   2. Purify the amplification product using any PCR product purification kit, and then recover the fragment in 20 μl of milliQ water.
   3. Set up a 50 μl PCR reaction using a mixture of the previously purified PCR products (0.5 μl from each) as a template and the primers A + B.
   4. Purify the amplification product.
   5. Clone the purified PCR product into a suitable cloning vector such as pGEM-T easy or pCR-TOPO.
   6. Sequence the obtained clones to confirm the correct assembly of the amiRNA into the precursor backbone (see Note 1).
   7. Subclone the amiRNA precursor behind a promoter of interest into an appropriate binary plasmid (see Note 2).
   8. Once cloned into an entry vector, a wide variety of binary destination vectors containing different promoter can be chosen. We often use the 35S promoter for strong constitutive expression of amiRNAs, but tissue-specific and inducible promoters have been tested successfully as well. It is important to mention that stronger promoters can

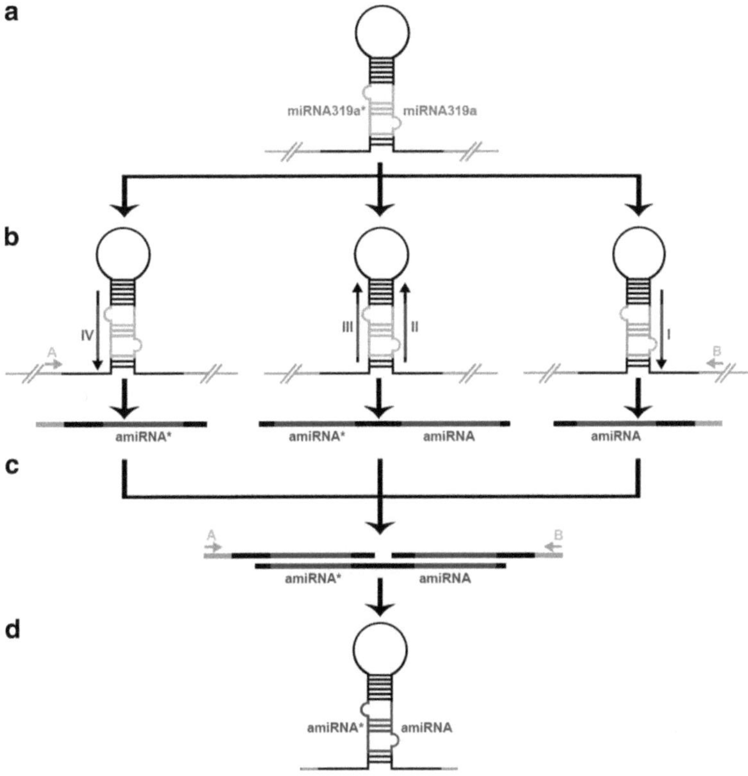

Fig. 1. Cloning amiRNA into miRNA319a precursor backbone. Schematic representation of the cloning procedure used to engineer constructs containing an amiRNA into the *Arabidopsis* miRNA319a precursor. (**a**) Representation of the miRNA319a precursor cloned into the pRS300 vector. The *light gray* areas represent the sequences of the miRNA/miRNA*; *black lines* correspond to the precursor sequence. (**b**) Use primers A, B, I, II, III, IV to set up three PCR reactions with the miRNA319a precursor as template. (**c**) Set up a second PCR reaction using primers A+B and a mixture of the purified products as template. (**d**) Clone the final PCR product containing the engineered amiRNA into a suitable vector. *Dark gray* lines represent the amiRNA sequence. *Labeled arrows* represent the different primers needed for the reactions.

induce more dramatic effects than weaker ones. Therefore, we do not recommend using weak promoters when strongly expressed genes should be silenced efficiently. They might nevertheless become very useful when only partial silencing is intended.

9. Introduction of the amiRNA precursor constructs in *A. thaliana* can be achieved by floral dipping (14) after generating an *Agrobacterium tumefaciens* strain bearing the binary construct (see Note 3).

3.3. AmiRNA Expressing Plants

1. Primary transformants should be selected with appropriate selection markers. As observed for any transgenic plant, the expression levels of the amiRNA can vary in different transgenic lines even when the same promoter is used. This is

mainly because of different insertional positions in the genome. Therefore, primary transformants may vary in their phenotypes. However, the lack of a phenotype can have many different explanations. The phenotype might not be detectable under the plant growth conditions, downregulation of the target gene might not be sufficiently strong to detect phenotypic changes, or redundancy in a gene family could hide the phenotypic effect of gene silencing.

2. Our results suggest that around 75% of amiRNAs designed with the current protocol lead to efficient silencing of the target gene in *Arabidopsis*. However, we recommend generating two amiRNAs per target gene with target sites in different regions of the target transcripts simultaneously to increase the chances of success.

3. In our experience, expression levels of the amiRNAs and their capacity to silence the target gene are stable in further generations of transgenic plants.

3.4. Confirming the amiRNA Functionality

1. Once the stable transgenic lines are obtained, it is necessary to confirm the amiRNA expression level, its processibility, and repression and cleavage of the target gene.

2. As a first step in the characterization of single insertion transgenic plants, quantitative PCR (qPCR) might be performed to establish the expression level of the amiRNA precursor. A primer for the miRNA precursor and a specific primer for the engineered amiRNA or amiRNA* are used to determine the amiRNA expression levels. However, high expression levels of a given miRNA precursor may not necessarily mean high levels of the mature miRNA. Therefore, a northern blot analysis must be done to confirm the correct processing of the amiRNA from its precursor.

3. Once the expression and processing of the amiRNA transgenic lines has been established, it is then necessary to determine whether the amiRNA is able to successfully repress the target gene expression. Since miRNA-mediated gene silencing effects are mostly evident on the transcript level, the degree of target downregulation can be easily tested by RT–PCR using oligos spanning the amiRNA target site. However, like some endogenous miRNAs, an amiRNA might also function via translational inhibition. In this case, western blot analysis using the appropriate antibodies would detect the translational repression. If no clear phenotype, amiRNA expression or target cleavage are detected in the T1 lines, we recommend designing new alternative amiRNA. Note that the use of amiRNA to silence genes involved in the miRNA biogenesis pathway could give rise to feedback effects that are difficult to interpret.

4. It is important to highlight that cross-silencing of the designed amiRNA to closely related genes is not trivial to predict and careful selection of the target sequence during the amiRNA design is critical.
5. Usually it is also important to show that the target gene down-regulation is a consequence of the amiRNA-mediated target cleavage and is not achieved through a different silencing mechanism. It is important to understand that the degradation of the target gene by another silencing mechanism could generate several small RNA molecules from a single target gene and increase the chances of cross-silencing of unrelated genes. Analysis of the target mRNA truncated species can determine which mechanism is governing the gene silencing. Cleavage of the target mRNA is produced in one well-established position in the miRNA-mediated pathway, while several truncated species appear if the siRNA pathway is involved.
6. To determine the cleavage position of the target gene and detect possible secondary cleavage species, a modified version of the invitrogen GeneRacer Kit (L1502-01) can be used. This kit uses two first steps of RNA dephosphorylation and cap removal to generate RNA molecules for the RNA oligoligation. Dephosphorylation eliminates 5′-PO_4 molecules of the truncated RNAs, but not from the capped ones. This makes it posterior ligation to the RNA oligo impossible. Removal of the mRNA cap structure allows for RNA oligoligation and further amplification and sequencing. Both these steps are critical to elucidate 5′-ends of full-length mRNA reducing the adapter ligation to degradated RNA.

Since the amiRNA-mediated slicing of the target mRNA generates uncapped, truncated molecules, the original technique is suitable to detect these RNA molecules just avoiding the first two steps. Skipping the dephosphorylation step ensures that the truncated mRNA retain the 5′-PO_4 molecule allowing its posterior ligation to the RNA adapter. Skipping the decapping step eliminates the possibility that full-length mRNAs become bound to the RNA adapter and reduce the chances of detecting unsliced target mRNA.

An additional DNAse treatment step should be performed on the isolated RNA samples to avoid contamination with genomic DNA. A schematic representation of the protocol is shown in Fig. 2.

3.5. Phenotypic Complementation of amiRNA Expressing Plants

All of the above-described methodologies used to express amiRNAs and to confirm their functionalities are also applicable to the overexpression of endogenous miRNA. Overexpression or expression of an endogenous miRNA in a specific tissue/developmental stage could provide valuable information about its function. Either the complete genomic locus or the sequence of

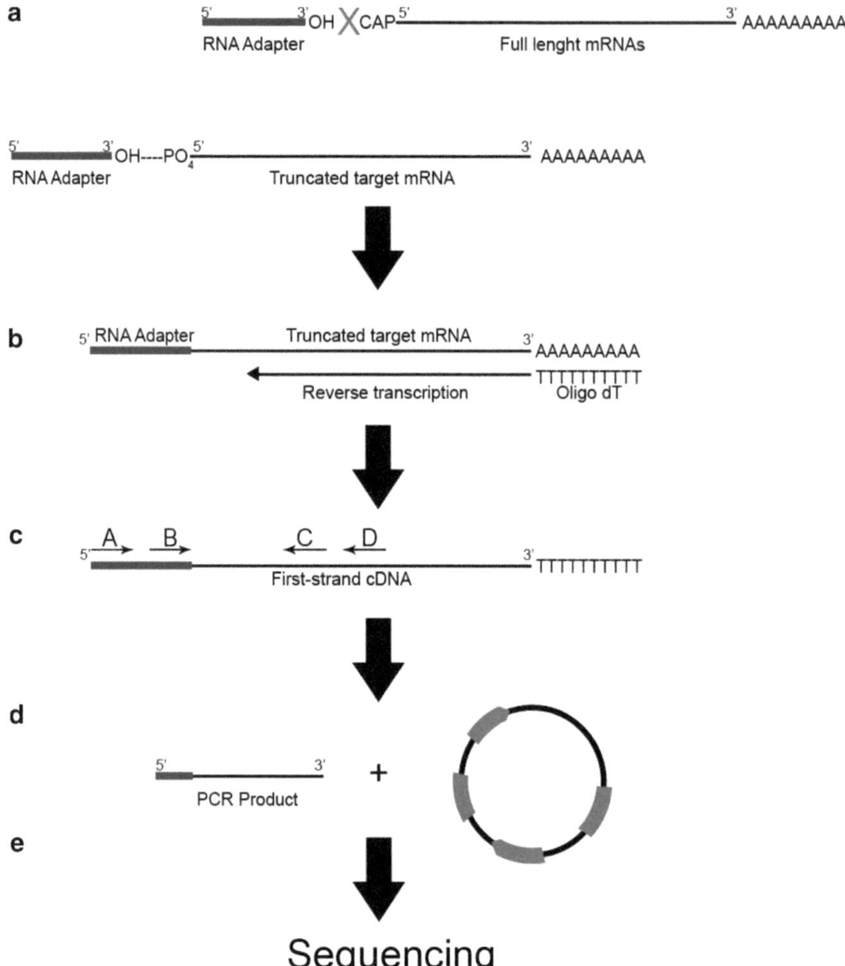

Fig. 2. AmiRNA-mediated target mRNAs slicing confirmation. Schematic representation of the modified 5′-RACE procedure used to determine cleavage site of the amiRNA over the target mRNA. (**a**) Ligation of an RNA adapter to the amiRNA-mediated truncated mRNA. Full-length mRNA is capped and will not ligate to the adapter. (**b**) Truncated mRNAs bound to the RNA adapter are reverse-transcribed using oligo dT to generate a first-strand cDNA. (**c**) A PCR reaction is set up using primers A (located in the adapter) and D (target gene-specific primer). A second reaction is set up using the purified PCR product as a template with nested primers (B in the adapter and C gene specific). (**d**) The purified PCR product is cloned in a suitable cloning vector. (**e**) Following confirmation that the construct has been inserted into the vector, the obtained clones are sequenced to identify the slicing site of the amiRNA.

the miRNA precursor can be cloned into an appropriate binary vector and transformed into plants as described above for amiRNAs.

Since the region where miRNAs bind to their target transcripts is small, it is possible to engineer silent mutations within the target site such that the transcript is no longer susceptible to amiRNA-mediated regulation. The use of these kinds of constructs as a control are fundamental to confirm that a given phenotype observed in the miRNA-expressing plants is due to the regulation of a single gene and is not achieved through the silencing of several genes (15). Re-introduction of the mutated transgene should complement the phenotype only if the phenotype is caused by amiRNA-mediated downregulation of the target gene. Silent mutations have been also successfully used to release regulation by endogenous miRNAs. Although a few mutations in the cleavage site (positions 10–11) might be enough to avoid the target gene cleavage, it has been shown that these kinds of mRNAs (with mutations in position 10–11) both endogenous and artificial can sequester the corresponding miRNA and prevent it to accomplish its functions in other genes (16). This phenomenon could result in misinterpretations of the phenotype. To avoid this problem, the silent mutation constructs should be engineered to contain as many mutations as possible.

4. Notes

1. It is important to highlight that the secondary structure of the miRNA precursors is only formed if the clone is in the sense orientation.
2. Many restriction sites are available in the flanking region of the miRNA319a precursor in the pRS300 plasmid. Alternatively, and recommended in most cases, the amiRNA precursor could be cloned into an appropriate gateway entry vector followed by recombination into binary vectors. If a gateway-based technology is chosen, the restriction digestion cloning step can be skipped and primers A and B with recombination sites in their sequences can be used.
3. When target genes are also available as transgenes behind strong promoters, both can be simultaneously and transiently introduced into tobacco leaves to test their functionality.

Acknowledgments

The authors would like to thank Proffesor Detlef Weigel and the microRNA team at his lab for continuous and helpful discussion and Beth Rowans for manuscript comments and text editing. Authors are supported by the European Community FP6 IP SIROCCO (contract LSHG-CT-2006-037900) and by the Max Planck Society.

References

1. Voinnet, O. (2009) Origin, biogenesis, and activity of plant microRNAs. *Cell.* **136**, 669–687.
2. Lanet, E., Delannoy, E., Sormani, R., Floris, M., Brodersen, P., Crété, P., Voinnet, O., Robaglia, C. (2009) Biochemical Evidence for Translational Repression by Arabidopsis MicroRNAs. *Plant Cell.* **21**, 1762–1768.
3. Schwab, R., Ossowski, S., Riester, M., Warthmann, N., and Weigel, D. (2006) Highly specific gene silencing by artificial microRNAs in Arabidopsis. *Plant Cell.* **18**, 1121–1133.
4. Ossowski, S., Schwab, R., and Weigel, D. (2008) Gene silencing in plants using artificial microRNAs and other small RNAs. *Plant J.* **53**, 674–690.
5. Warthmann, N., Chen, H., Ossowski, S., Weigel, D., and Herve, P. (2008) Highly specific gene silencing by artificial miRNAs in rice. *PLoS One.* **3**, e1829.
6. Khraiwesh, B., Ossowski, S., Weigel, D., Reski, R., and Frank, W. (2008) Specific gene silencing by artificial MicroRNAs in Physcomitrella patens: an alternative to targeted gene knockouts. *Plant Physiol.* **148**, 684–693.
7. Molnar, A., Bassett, A., Thuenemann, E., Schwach, F., Karkare, S., Ossowski, S., et al. (2009) Highly specific gene silencing by artificial microRNAs in the unicellular alga *Chlamydomonas reinhardtii. Plant J.* **58**, 165–174.
8. Niu, Q.W., Lin, S.S., Reyes, J.L., Chen, K.C., Wu, H.W., Yeh, S.D., et al. (2006) Expression of artificial microRNAs in transgenic Arabidopsis thaliana confers virus resistance. *Nat. Biotech.* **24**, 1420–1428.
9. Duan, C.G., Wang, C.H., Fang, R.X., and Guo, H.S. (2008) Artificial MicroRNAs Highly Accessible to Targets Confer Efficient Virus Resistance in Plants. *J. Virol.* **82**, 11084–11095.
10. Alvarez, J.P., Pekker, I., Goldshmidt, A., Blum, E., Amsellem, Z., and Eshed, Y. (2006) Endogenous and Synthetic MicroRNAs Stimulate Simultaneous, Efficient, and Localized Regulation of Multiple Targets in Diverse Species. *Plant cell.* **18**, 1134–1151.
11. Montgomery, T.A., Howell, M.D., Cuperus, J.T., Li, D., Hansen, J.E., Alexander, A.L., et al. (2008) Specificity of ARGONAUTE7-miR390 interaction and dual functionality in TAS3 trans-acting siRNA formation. *Cell.* **133**, 128–141.
12. Khvorova, A., Reynolds, A., and Jayasena, S.D. (2003) Functional siRNAs and miRNAs exhibit strand bias. *Cell.* **115**, 209–216.
13. Schwarz, D.S., Hutvagner, G., Du, T., Xu, Z., Aronin, N., and Zamore, P.D. (2003) Asymmetry in the assembly of the RNAi enzyme complex. *Cell.* **115**, 199–208.
14. Weigel, D., and Glazebrook, J. (2002) Arabidopsis: A Laboratory Manual. Cold Spring Harbor, NY: Cold Spring Harbor Laboratory Press. 354 p
15. Palatnik, J.F., Allen, E., Wu, X., Schommer, C., Schwab, R., Carrington, J.C., et al. (2003) Control of leaf morphogenesis by microRNAs. *Nature.* **425**, 257–263.
16. Franco-Zorrilla, J.M., Valli, A., Tudesco, M., Mateos, I., Puga, M.I., Rubio-Somoza, I., et al. (2007) Target mimicry provides a new mechanism for regulation of microRNA activity. *Nat Genet* **39**, 1033–1037.

Chapter 10

Mimicry Technology: Suppressing Small RNA Activity in Plants

Ignacio Rubio-Somoza and Pablo Andrés Manavella

Abstract

Small RNA suppression constitutes one of the major difficulties for a full molecular characterization of their specific roles in plants. Taking advantage of the latest insights into the new post-biogenesis layer of regulation in microRNA (miRNA) activity, it is possible to overcome the above-mentioned limitation (Nat Genet 39:1033–1037, 2007). We engineered the IPS1 non-coding RNA to bear a complementary sequence to a given miRNA family, resulting in specific sequestration of RISC complexes. MIMIC technology allows for the constitutive release of all of the potential targets of a miRNA family as well as tissue-specific and inducible suppression of its activity.

Key words: Target mimicry, microRNA, Small RNA suppression, IPS1

1. Introduction

IPS1 encodes a non-coding RNA that bears a region that is nearly exactly complementary to miRNA399, and both are involved in the phosphate starvation response. Complementarity of both sequences is disrupted by three nucleotides in the IPS1 transcript at the predicted miRNA399 cleavage site, resulting in the specific sequestration of $RISC^{399}$ complexes. Therefore, IPS1 expression releases the natural miRNA399 targets from repression, allowing them to be translated into proteins (Fig. 1). IPS1 belongs to a small and evolutionarily conserved family of ncRNAs with a common potential role in miRNA399 regulation. For the first time, we have successfully engineered the IPS1 sequence by replacing the miR399 target sequence with miR156 and miR319 and constitutively suppressed the activity of these two microRNA (miRNAs) in *Arabidopsis thaliana* (1). Further validation of this technique was achieved by targeting other Arabidopsis miRNA families. MIMIC technology

Tamas Dalmay (ed.), *MicroRNAs in Development: Methods and Protocols*, Methods in Molecular Biology, vol. 732, DOI 10.1007/978-1-61779-083-6_10, © Springer Science+Business Media, LLC 2011

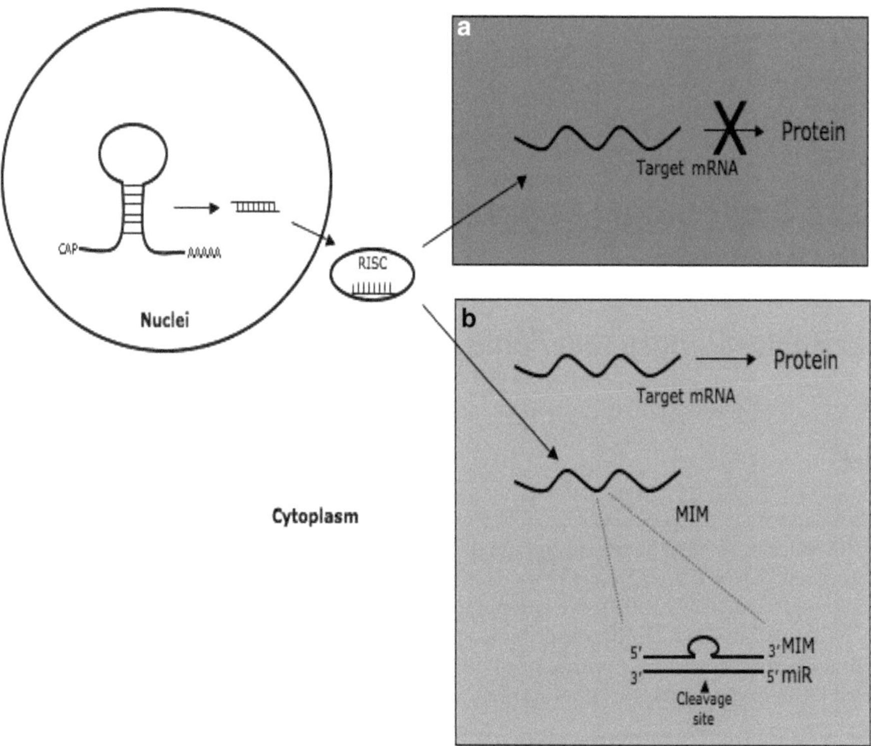

Fig. 1. *MIMIC technology releases natural sRNA targets from this level of regulation by specifically sequestering RISC loaded complex*. (**a**) sRNAs, miRNAs in this case, are loaded into the RISC complex that scans cell cytoplasm for mRNAs showing highly complementary sequences. miRNA targeting results in mRNA translation suppression due to cleavage or retention of the transcripts. (**b**) Presence of a MIMIC transcript bearing a specific complementary stretch resembling natural target sequence results in the specific sequestration of RISC complexes, allowing the transcripts of natural targets to translate into proteins. Disruption of the miRNA-MIMIC pairing at the predicted cleavage position is the main feature that allows for RISC retention and release of miRNA targets.

can also be used in different plant species and is predicted to work with two mechanisms of mRNA regulation by small RNAs (sRNAs): transcript cleavage and translational inhibition.

Next, we will cover the key points for a successful MIMIC design and provide some advice for a systematic sRNA study. Building MIMIC constructs relies on the replacement of miRNA399 target sequence within IPS1 by a two-step PCR. Selection of a suitable promoter to drive MIMIC expression is one of the major considerations that needs to be taken in account for a detailed study of sRNA function.

2. Material

2.1. Building MIMIC Constructs and Cloning

1. Commercial kit to isolate genomic DNA PCR quality from *A. thaliana* seedlings or leaves.

2. PCR reagents (specific primers detailed below, Fusion DNA polymerase (NEB), dNTPS) and performing device, Thermocycler.
3. Equipment to perform DNA electrophoretic assays (BioRad), Agarose, TAE buffer and Kit for DNA isolation from agarose gels (Wizard, PROMEGA).
4. PCR acceptor vector (pGEM-Teasy (PROMEGA) or PCR-GW TOPO (INVITROGEN)).
5. Binary vector with specific promoter to drive MIMIC expression.
6. *Escherichia coli* and *Agrobacterium tumefaciens* competent cells for plasmid propagation and plant transformation.

3. Methods

3.1. Primer Design for Specific miRNA Attenuation

One of the main advantages of the MIMIC technology is the ability to suppress an entire miRNA family due to the high sequence homology shared between its members. However, some large miRNA families can show small differences among groups of their members (e.g., miRNA169 family). Design of individual MIMIC constructs mimicking sequence targets for those different groups is recommended (i.e., MIM169ABCD and MIM169EFGH, for miRNA169ABCD and EFGH family members, respectively).

MiRNA sequences from *A. thaliana* and other plant organisms can be obtained from the publicly available databases miRBase (http://www.mirbase.org/) and ASRP (asrp.cgrb.oregonstate.edu). MIMIC technology is also predicted to be functional for miRNA and sRNA sequences different from the ones described in the above-mentioned databases.

Replacement of the miRNA399 target within the IPS1 gene by the custom desired MIMIC sequence is the first step. MIMIC sequence is the reverse complement of the corresponding sRNA to attenuate in which a three-nucleotide bulge must be introduced after the predicted cleavage position (normally between the bases in the 10th and 11th positions from the miRNA 5′ end). The bulge sequence must not contain a possible cleavage reconstitution site in order to be functional.

A useful tool for obtaining the complementary reverse sequences of the sRNAs under study is the EMBOSS REVSEQ tool (http://emboss.bioinformatics.nl/cgi-bin/emboss/revseq).

1. Paste the sRNA sequence in the indicated place on the website, then follow the instructions to generate the reverse complementary sequence.

2. Take this sequence and insert three nucleotides after the nucleotide in the 10th position from the 3′ outermost nucleotide (that corresponds to sRNA 5′ end). An important parameter to ensure MIMIC efficiency is base-bulge composition. It is strongly recommended to avoid using nucleotides within the bulge that are the same as those flanking the 10th and 11th positions (see Note 1).
3. Design the MIMIC forward primer (MIM.fw) by taking the resulting sequence from the former step and pasting it into the position indicated with the underlined X in the following sequence:
MIM.fw: 5′-CTAGAAA*X*AGCTTCGGTTCCCC-3′
4. Design the MIMIC reverse primer (MIM.rev) by replacing the underlined X in the following sequence with the reverse complementary sequence from step 2.
MIM.rev: 5′-GGGGAACCGAAGCT*X*TTTCTAGAGG-3′

3.2. IPS1 Gene Amplification

1. Set up a PCR using 1 μl of *A. thaliana* genomic DNA (200 ng) as the template in a final volume of 25 μl. Use the following primers for this PCR:
   (a) IPS1.fw: 5′-AAAACACCACAAAAACAAAAG-3′
   (b) IPS1.rev: 5′-AAGAGGAATTCACTATAAAGAG -3′
2. Load and run the product of the PCR in an agarose-TAE gel (1% w/v) along with a DNA size marker. The expected band is 542 bp long (see Note 2).
3. Isolate the band from the gel using a scalpel under UV light. Purify the DNA using a commercially available kit (i.e., Wizard SV gel and PCR cleanup system, PROMEGA) and follow the manufacturer's instructions.
4. Clone the isolated DNA into a plasmid (i.e., pGEM-T easy, PROMEGA, or pCR8/GW/TOPO TA, INVITROGEN) to verify the sequence and use it as a PCR template for the following steps (see Note 3).

3.3. MiRNA Target Sequence Replacement in IPS1 Coding Sequence

1. Set up two independent PCRs using 1 μl of the plasmid containing IPS1 coding sequence as the template in a 25 μl final reaction volume and use the conditions described in step 1 of the previous section.

   Use the following primer combination in PCR1 to amplify the 5′ region:

   IPS1.fw: 5′-AAAACACCACAAAAACAAAAG-3′

   MIM.rev: 5′-GGGGAACCGAAGCT*X*TTTCTAGAGG-3′

   Use the following primer combination to generate the 3′ region in PCR2:

MIM.fw: 5′-CTAGAAA*X*AGCTTCGGTTCCCC-3′

IPS1.rev: 5′-AAGAGGAATTCACTATAAAGAG -3′

2. Load and run the products of both PCRs in separate slots in an agarose-TAE gel (1% w/v) along with a DNA size marker. The expected size for the products from PCR1 and PCR2 are 314 and 273 bp, respectively.
3. Isolate the bands separately in two Eppendorf tubes and use a kit to extract DNA (using the same procedure that is described in step 3 of the previous section).
4. Set up a final PCR in a single tube using 1 μl of each of the products obtained from PCR1 and PCR2 as templates and IPS1.fw and IPS1.rev as primers.
5. Load and run the products of this PCR in an agarose-TAE gel (1% w/v). Expected size of the final PCR product is 542 bp.
6. Isolate the correct band of DNA from gel by following the same procedure described above.
7. Proceed to clone MIMIC in a vector with a comprehensive multi-cloning site (MCS), i.e., PGEM (PROMEGA), or in a Gateway DONOR vector (pCR8/GW/TOPO TA, INVITROGEN) to verify the sequence, and then subclone into a final binary vector for plant transformation under the control of the desired promoter.

3.4. Promoter Election to Drive MIMIC Expression

A plethora of vectors with constitutive (viral 35S and plant ubiquitin derived), inducible, or tissue- and cell-specific promoters and compatible with the comprehensive MCS and GATEWAY systems is currently publicly available. In our experience, the virus-derived 35S promoter guarantees a constitutive expression pattern. Nevertheless, constitutive sequestration of sRNAs affecting reproductive and seed viability traits might lead to the recovery of Arabidopsis lines with low expression levels of the MIMIC construct that exhibit remnant sRNA activity as a consequence. An ethanol-inducible system has been also successfully used in our laboratory to drive MIMIC expression. Organ-specific promoters such as Apetala3, which is expressed only in floral organs, and Clavata 3 (CLV3), which is expressed only in meristematic domains, have also been successfully employed in our laboratory (2). Among other sources of binary vectors, the Weigel and Schmid research groups have developed a large collection of traditional and gateway compatible plasmids with constitutive, tissue- and cell-specific promoters, which is publicly available by request (weigel@tuebingen.mpg.de, markus.schmid@tuebingen.mpg.de).

The ASRP and MPSS (mpss.udel.edu/at/) publicly available databases include expression data from several sRNA libraries from different Arabidopsis tissues, and can be used as a guide to choose the best promoter to drive MIMIC constructs.

Introduction of MIMIC constructs in *A. thaliana* can be achieved by floral dipping (3) after generating an *A. tumefaciens* strain bearing the binary construct from step 7.

3.5. Impact of MIMIC Technology on sRNA Target Transcripts

Whether MMIC technology has successfully attenuated sRNA activity and the consequent regulatory release of mRNA targets can be addressed using several approaches.

Quantitative real-time PCR (QRT-PCR) with specific primers that flank the predicted target site on the released transcripts can be used to detect the expected increase in the levels of expression of these targets. However, only the role of cleavage-mediated sRNA regulation can be studied with this approach, since inhibition of translation is not reported to work through mRNA degradation in plants. Ideally, an increase in protein levels of the target genes detected by western blot would be the definitive confirmation of MIMIC success. Additional ways to partially corroborate increase in target protein levels would come from the generation of a translational fusion of the corresponding genomic locus fused to reporter genes, such as GFP or GUS, and compare the reporter expression in presence and absence of the MIMIC transcript.

4. Notes

1. miRNA-dependent target slicing normally takes place between positions 10 and 11 from the outermost 5′ base of the miRNA. We suggest to previously confirm the cleavage site in the available literature or by performing 5′ RACE on predicted miRNA targets. In case of shifting in the main cleavage site, the three-nucleotide bulge must be relocated within accordingly complementary IPS1 target site.
2. We have also successfully engineered a smaller IPS1 transcript encompassing the 50 bs flanking the mimic site, using a 124 nucleotide decoy. This smaller version mirrors the developmental defects obtained when a whole IPS1 backbone is used.
3. PCR plasmid acceptors are fashioned on the ability of Taq-polymerases to add an overhang 5′-A on the amplification products. In case of using a different PCR polymerase than Taq, a reaction to add an overhang 5′-A is needed before to tackle the ligation with the acceptor plasmid.

Acknowledgments

The authors thank Prof. Detlef Weigel and the microRNA team at his lab for continuous and helpful discussion and Beth Rowans for manuscript comments and text editing. Authors are supported by European Community FP6 IP SIROCCO (contract LSHG-CT-2006-037900) and by the Max Planck Society.

References

1. Franco-Zorrilla, J.M., Valli, A., Todesco, M., Mateos, I., Puga, M.I., Rubio-Somoza, I., et al. (2007) Target mimicry provides a new mechanism for regulation of microRNA activity. *Nat Genet* **39**, 1033–1037.
2. Brand, U., Grunewald, M., and Simon, R. (2002) regulation of CLV3 expression by two homeobox genes in arabidopsis. *Plant Phys* **129**, 565–575.
3. Weigel, D., and Glazebrook, J. (2002) Arabidopsis: A Laboratory Manual. Cold Spring Harbor, NY: Cold Spring Harbor Laboratory Press. 354 p

Chapter 11

Experimental Validation of MicroRNA Targets Using a Luciferase Reporter System

Francisco E. Nicolas

Abstract

MicroRNAs (miRNAs) are a class of small noncoding transcripts that repress gene expression by pairing with their target messenger RNAs (mRNAs). The human genome codes for hundreds of different miRNAs and it is predicted that they target thousands of mRNAs involved in a wide variety of physiological processes such as development and cell identity. In animals, the identification of mRNA targets is complex because most miRNAs and their target mRNAs do not have exact or nearly exact complementarity. This tendency of animal miRNAs to bind their mRNA targets with imperfect sequence homology represents a considerable challenge to identifying miRNA targets. Computational algorithms based on conservation and experimental approaches based on expression profiles are flooding the literature with lists of candidate genes containing a large number of false-positive and false-negative predictions and indirect targets that cover the real list of direct targets for each miRNA. Currently, the only available tools to validate a sequence as a direct target of an miRNA are the systems based on a reporter gene carrying the candidate sequence. Here, an miRNA target validation reporter gene system based on the luminescence generated by the luciferase protein is described in detail, including the design of the reporter constructs, its expression in a model cell line and its measurement using a luciferase assay.

Key words: miRNA, miRNA target, Target validation, miRNA mimics, Luminescence, Luciferase assay

1. Introduction

MicroRNAs (miRNAs) are endogenous, single-stranded, short (19–21 nt) RNA molecules that regulate the expression of protein-coding genes (1). These RNA molecules fine-tune the level of proteins in complex and overlapping patterns rather than completely switching off the expression of their target genes (2–4). Although the effects of this fine-tuning might be small, the large number of genes regulated by miRNAs suggests that this is a significant regulatory layer. It is currently accepted that they play an important role in many complex processes such as embryo

Tamas Dalmay (ed.), *MicroRNAs in Development: Methods and Protocols*, Methods in Molecular Biology, vol. 732, DOI 10.1007/978-1-61779-083-6_11, © Springer Science+Business Media, LLC 2011

development, tissue identity, proliferation, differentiation, apoptosis, signalling pathways, metabolism, cancer, and viral infections (5). The first step in the miRNA biogenesis is the synthesis of a long primary transcript (pri-miRNA) that folds back on itself to form a distinctive hairpin structure (4). This long pri-miRNA is first recognised and cleaved by Drosha, a ribonuclease III (RNase III), which liberates a 60–70 nt stem–loop intermediate known as the miRNA precursor (pre-miRNA) (6, 7). The pre-miRNA resulting from this first cleavage is then actively transported across the nuclear membrane to the cytoplasm by Ran-GTP and the export receptor exportin-5 (8, 9). Once in the cytoplasm, a second cleavage event is carried out by Dicer, another RNase III endonuclease, which releases the loop and produces the miRNA duplex (10, 11). After cleavage, the RNA duplex is incorporated into the effector complex RISC (RNA-induced silencing complex). During this loading into the RISC, the RNA duplex that consists of the guide strand (or mature miRNA) and its complementary strand (passenger strand or miRNA*) is unwound (12–14). The mature miRNA is incorporated into the RISC and the miRNA* is degraded. However, the passenger strand also has a chance of being loaded into the RISC and, therefore, could be used in gene regulation. In animals, mature miRNAs guide the RISC to target messenger RNAs (mRNAs) that contain sequences partially complementary to the miRNAs, downregulating gene expression by either translational repression or mRNA decay (15–19) (Fig. 1). Finding the targets of miRNAs is easy, if they complement each other, since several genome sequences have already been completed. This is generally true in plants, where many targets can be

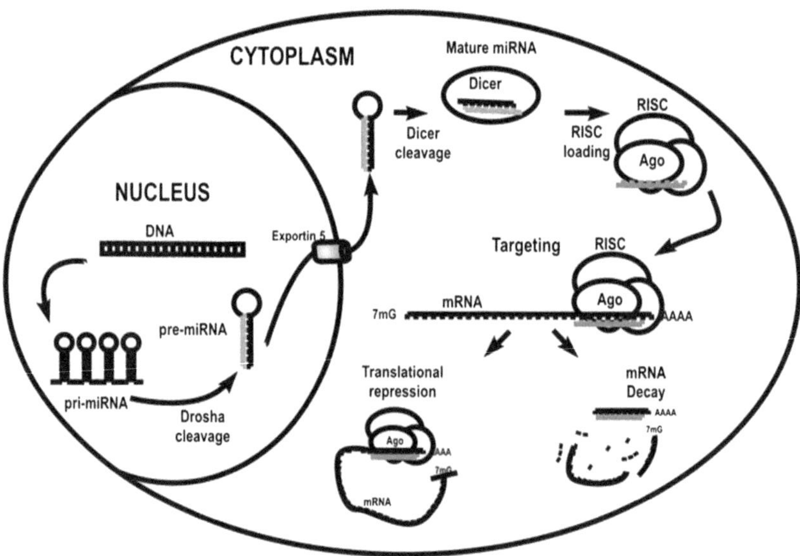

Fig. 1. MiRNA biogenesis pathway.

identified simply by searching for complementary regions to the miRNAs (20). However, animal miRNAs are not perfectly complementary to their targets with few highly complementary to their target mRNA (21, 22). Animal miRNAs recognise their target mRNAs through the so-called seed sequence, which is the 2–8 nt sequence at the 5′ region of the miRNA. This 5′ region is the most conserved portion of the metazoan miRNAs (23). Two approaches are used to identify candidate genes targeted by miRNAs: computational target prediction algorithms and experimental target identification strategies. Target prediction algorithms are diverse and use different parameters to provide candidate target genes for miRNAs (24). The main weakness of these computational programs is that different algorithms predict different lists of putative targets for a given miRNA (25), which suggests that there are both false-positive and false-negative predictions. Nevertheless, prediction tools are useful for identifying potential targets that can be validated experimentally afterwards (26). Experimental approaches can be subdivided into five different strategies: miRNA overexpression using siRNAs mimics (27), miRNA downregulation using antagomirs or antimirs (25, 28), the combined overexpression and downregulation of miRNAs (29), and RISC purification using antibodies against Argonaute-2 (30, 31). In all these four cases, the mRNA is purified after treatment and analysed using microarray hybridisation or sequencing. The fifth strategy is based on the main regulatory action of miRNAs occurring at the translational level, reducing the amount of translated protein from the target mRNA. This strategy consists of detecting changes in protein levels after the misregulation of miRNA activity. The strategy has been proposed in two recent studies that suggested using stable isotope labelling of amino acids in cell culture, a technique applied for high-throughput protein profiling, to investigate the effect of miRNAs on the cell proteome (32, 33).

The main problem of all these strategies is that they only produce a list of candidate genes that could be either a direct target of the tested miRNA, when the reduction in the mRNA or protein level of the target is because of the action of the targeting miRNA on the expression of the target, or an indirect target, when the miRNA is targeting a secondary unknown gene responsible for the downregulation of the first indirect target. The only experimental assays capable of validating a candidate gene as a direct target of an miRNA are those based on the differential activity of a reporter gene attached to a candidate sequence, when this quimeric gene is expressed in the presence or absence of the targeting miRNA. This chapter describes the design and construction of the necessary plasmids, culture, and transfection into a model cell line and the performance of the luminescence assay for a reporter system based on the activity of the luciferase gene.

2. Materials

2.1. DNA Constructs

2.1.1. PCR DNA Amplification

1. DNA-specific oligonucleotides for PCR amplification (Sigma–Aldrich).
2. Phusion® High-Fidelity DNA Polymerase (F-530) (Finnzymes Oy).
3. GoTaq DNA polymerase (Promega).
4. Solution of $MgCl_2$: 50 mM.
5. PCR nucleotide mix: ready-to-use premixed solution of PCR-grade dATP, dCTP, dGTP, and dTTP (10 mM each) (Roche).

2.1.2. DNA Cloning

1. pGem-T easy vector system (Promega).
2. pGL3-enhancer vector (Promega).
3. T4 DNA ligase (Roche).
4. Restriction enzymes: *Xba*I, *Nhe*I, and *Sac*I (Roche).
5. *Escherichia coli* cloning system: DH5α heat shock competent cells (Invitrogen), LB medium (10 g/l tryptone, 5 g/l yeast extract, and 10 g NaCl), IPTG, X-gal, and ampicillin (Sigma).
6. 10× Annealing buffer: 100 mM Tris–glycine, pH 7.5, 1 M NaCl, and 10 mM EDTA.

2.2. Cell Culture and Transfection

1. Dulbecco's modified Eagle's medium (DMEM) (Gibco BRL) supplemented with 10% foetal bovine serum (Gibco BRL), 2 mM l-glutamine (200 mM stock, Gibco BRL), and 1× penicillin–streptomycin (100× stock, Gibco BRL).
2. OPTI-MEM I + GlutaMAX I medium (Gibco BRL).
3. Solution of 0.25% trypsin and 1 mM EDTA (Gibco BRL).
4. Phosphate-buffered saline (PBS) (Dulbecco A, Oxoid).
5. Lipofectamine 2000 (Invitrogen).
6. AllStars Negative Control siRNA and miScript miRNA Mimic (Qiagen).

2.3. Lysis and Luciferase Assay

1. Cell lysis buffer: 1% (v/v) Triton X-100, 25 mM glycylglycine pH 7.8, 15 mM $MgSO_4$, 4 mM EGTA, and 1 mM DTT added just before use. Alternatively: luciferase cell culture lysis reagent (Promega).
2. Luciferase assay substrate and luciferase assay buffer (luciferase assay system, Promega).

3. Methods

Reporter gene-based assays are currently the only available method to validate an mRNA as a direct target of a particular miRNA. Among these reporter assays, the luciferase-based vectors are the prevalent choice to assess the regulatory effects of a specific miRNA on its potential targets. These vectors usually contain a luciferase gene, typically from *Renilla* or firefly, under the transcriptional control of an upstream constitutive promoter and a downstream poly(A) signal. The candidate target sequence is inserted between the luciferase coding sequence and the poly(A) signal to transcribe a chimeric mRNA containing the luciferase-coding sequence and the candidate target sequence. After the transfection of this vector into a model cell line, luciferase expression is regulated by binding the targeting miRNA to the candidate sequence, which can be monitored by measuring the luciferase activity with a luminescence assay.

3.1. Preparation of Plasmids for miRNA Target Validation

The validation of the candidate sequence of an miRNA target requires two essential plasmids: a "wild-type" plasmid that contains the original 3′-UTR harbouring the candidate target sequence and a "mutant" plasmid that contains the same 3′-UTR but containing a mutated seed sequence in the candidate target sequence. These two plasmids derive from a basal cloning plasmid containing all the necessary elements to perform the luciferase assay: a constitutive promoter followed by a luciferase gene, a multiple cloning region where the candidate 3′-UTRs are inserted and a poly(A) signal to terminate the transcription of this chimeric mRNA. Finally, a positive control plasmid, which contains a perfect match sequence for the targeting miRNA, is necessary to verify the miRNA activity and the efficiency of the whole luciferase assay (Fig. 2):

1. Basal cloning plasmid: this plasmid can be engineered from any original plasmid designed to monitor luciferase activity such as, for instance, the pGL3 plasmid. A basal cloning plasmid derived from pGL3 is generated by first deleting the region between *Sac*I and *Bgl*II upstream of the SV40 promoter and then incorporating a multiple cloning site (MCS) (5′-TCTAGAGAGCTCAGATCTCCCGGGCTCG-AGGCTAGCCTCTAGA-3′) inserted into the *Xba*I site downstream of the luciferase stop codon. This MCS contains the *Sac*I and *Nhe*I restriction sites where the candidate target sequence is inserted to construct the wild-type and mutant plasmid.
2. Wild-type plasmid: first, the candidate target sequence is PCR amplified from genomic DNA using two specific primers and

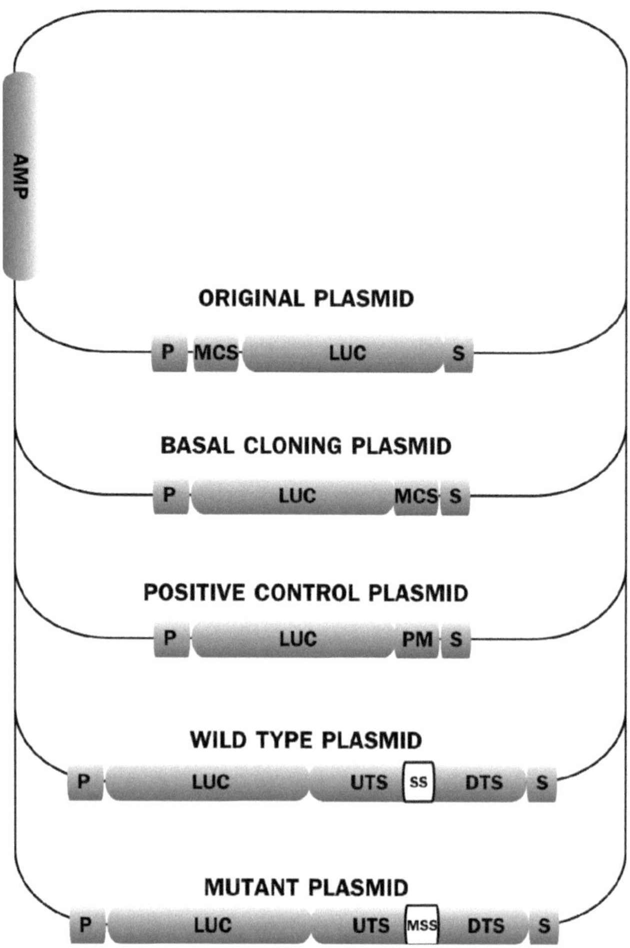

Fig. 2. Plasmids required for miRNA target validation. *P* promoter, *MCS* multicloning site, *LUC* luciferase gene, *S* transcription stop (poly(A) signal), *PM* miRNA perfect match target sequence, *UTS* upstream miRNA target sequence, *DTS* downstream miRNA target sequence, *SS* seed target sequence, *MSS* mutant seed target sequence, and *AMP* ampicillin resistance gene.

cloned into a T-hanging cloning plasmid such as, for example, pGem-T easy (Promega) (see Note 1). The fragment amplified has to be a minimum of 400, 200 bp upstream and downstream of the target sequence, to ensure the original folding structure of the target and the flanking sequences, which could affect the targeting efficiency of the miRNAs. Usually, the amplified sequence has a maximum 1 kb length, since longer sequences could interfere with the expression of the luciferase gene. Second, the fragment is PCR amplified from pGem-T using another two specific primers containing the *Sac*I and *Nhe*I restriction sites to the 5′ and 3′ ends, respectively, and inserted into the MCS of the basal plasmid.

3. Mutant plasmid: the candidate target sequence inserted in the wild-type plasmid is used as a template to generate the mutant

version using an overlapping strategy. First, two fragments are amplified from the wild-type plasmid using the primer pairs Forward1–Reverse1 and Forward2–Reverse2. The primers Forward1 and Reverse2 are the same primers designed to amplify the wild-type fragment from pGem-T and contain the *Sac*I and *Nhe*I restriction sites, respectively. The primers Reverse1 and Forward2 are complementary to each other and contain between four and six mismatches located at the position where the seed sequence of the targeting miRNA recognises the target sequence (see Note 2). The two fragments amplified with these two pairs of primers share a small overlapping region where the new mutated target sequence is located (see Note 3). In a second PCR, the two fragments are mixed equally and amplified using the primer pair Forward1–Reverse2, generating the mutant version of the candidate target sequence that can be cloned again into pGem-T easy (see Note 1). Finally, this mutant sequence is digested with *Sac*I and *Nhe*I and then inserted into the MCS of the basal plasmid to generate the mutant plasmid (Fig. 3).

4. Positive control plasmid: this plasmid is engineered using a pair of primers complementary to each other and containing the perfect match sequence to the targeting miRNA plus the *Sac*I and *Nhe*I restriction sites to the 5′ and 3′ ends, respectively. The two primers are annealed to each other in annealing buffer and inserted into the MCS of the basal plasmid to generate the positive control plasmid.

3.2. Transfection of NIH 3T3 Cells for miRNA Target Validation

NIH 3T3 is a mouse fibroblast cell line easily transfectable and unproblematic for luciferase expression (see Note 4). Cells are transfected using a transfection reagent suitable for both DNA plasmids and small RNAs (for instance, lipofectamine 2000, Invitrogen) with the basal, wild-type, mutant, or positive control plasmids, a synthetic siRNA that mimics the duplex generated by Dicer (see Subheading 1) or a scramble siRNA as a negative control. Lipofectamine only treated cells serve as an extra negative control for the luciferase activity. Transfections are carried out six times in triplicate using two independent plasmid preparations:

1. Plasmid preparations for transfections: the quality of the DNA to be transfected is important for obtaining reliable and reproducible luminescence values. The plasmid preparations have to be highly concentrated and free of impurities to increase transfection efficiency. This high quality can be achieved using a commercial column based on affinity resin (for instance, Qiagen Plasmid Midi Kit) (see Note 5).
2. The day before transfection, seed 2.5×10^4 cells per well of a 24-well plate in 500 μl of supplemented DMEM without antibiotic and incubate the cells at 37°C and 5% CO_2 in an incubator. The wells should be 30–50% confluent on the day of transfection (see Note 6).

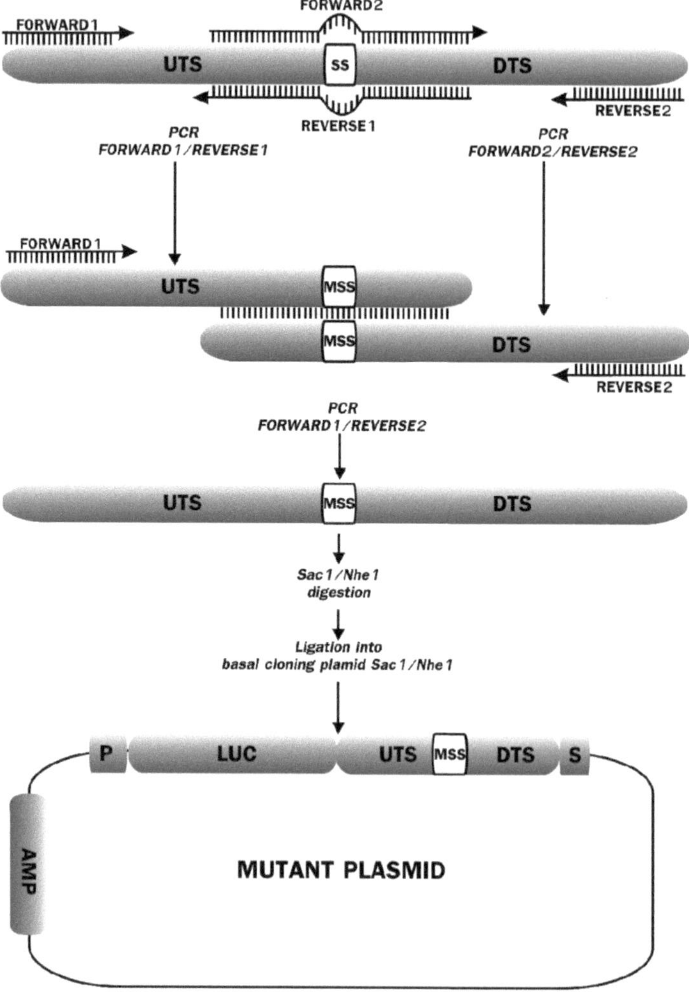

Fig. 3. Construction of the mutant plasmid for miRNA target validation. *P* promoter, *LUC* luciferase gene, *S* transcription stop (poly(A) signal), *UTS* upstream miRNA target sequence, *DTS* downstream miRNA target sequence, *SS* seed target sequence, *MSS* mutant seed target sequence, and *AMP* ampicillin resistance gene.

3. On the day of transfection, prepare two tubes, A and B, for each well containing:

A	B
125 μl of OPTI-MEM + 400 ng of plasmid + 2.5–25 pmol of siRNA (see Note 7)	125 μl of OPTI-MEM + 2 μl of lipofectamine

and for each of the following treatments:

(a) Only the positive control plasmid.

(b) Only the wild-type plasmid.

(c) Only the mutant plasmid.
(d) The positive control plasmid plus mimicking siRNAs.
(e) The wild-type plasmid plus mimicking siRNAs.
(f) The mutant plasmid plus mimicking siRNAs.
(g) The positive control plasmid plus scramble siRNAs.
(h) The wild-type plasmid plus scramble siRNAs.
(i) The mutant plasmid plus scramble siRNAs.
(j) Only lipofectamine.

4. Mix vigorously and let stand for 5 min, then combine tube A with tube B, mix vigorously and incubate for 20 min at room temperature (10–25°C) to allow the transfection complex to form. The amount of transfection mixture for each well can be scaled up according to the number of replicates.
5. While the complex is forming, wash the cells once with 500 μl of PBS. Then, gently aspirate the PBS and immediately add the transfection mixture to the cells in the 24-well plates.
6. Incubate the cells with the transfection complexes at 37°C and 5% CO_2 for 5 h, and then replace the transfection mixture with supplemented DMEM without antibiotic.
7. Finally, incubate the transfected cells for 48 h after transfection to obtain maximal levels of gene expression in the luciferase assay.

3.3. Luciferase Assay for miRNA Target Validation

1. Aspirate the cell culture growth medium completely and wash the cells twice with 500 μl of PBS.
2. Remove the remaining PBS and add 150 μl of lysis buffer to each well. Scrape cells off the plate using a rubber policeman (see Note 8).
3. Transfer the lysate to a microcentrifuge tube and incubate at room temperature (10–25°C) for 10 min.
4. Remove the cellular debris by spinning the lysates in a microcentrifuge for 1 min at maximum speed.
5. Transfer the supernatant to a new microcentrifuge tube and start the luciferase reaction immediately.
6. Transfer 75 μl of lysate from each sample to a 96-well black plate (F Poly Sorp, NUNC) and start the reaction by adding 50 μl of luciferase substrate (luciferase assay reagent, Roche) at room temperature. Save the rest of the lysate on ice for a BCA protein assay (see below).
7. Measure the light emission within 10 s of adding the luciferase substrate (see Note 9) using a multilabel counter (for instance, Victor^2, Perkin-Elmer, MA, USA).

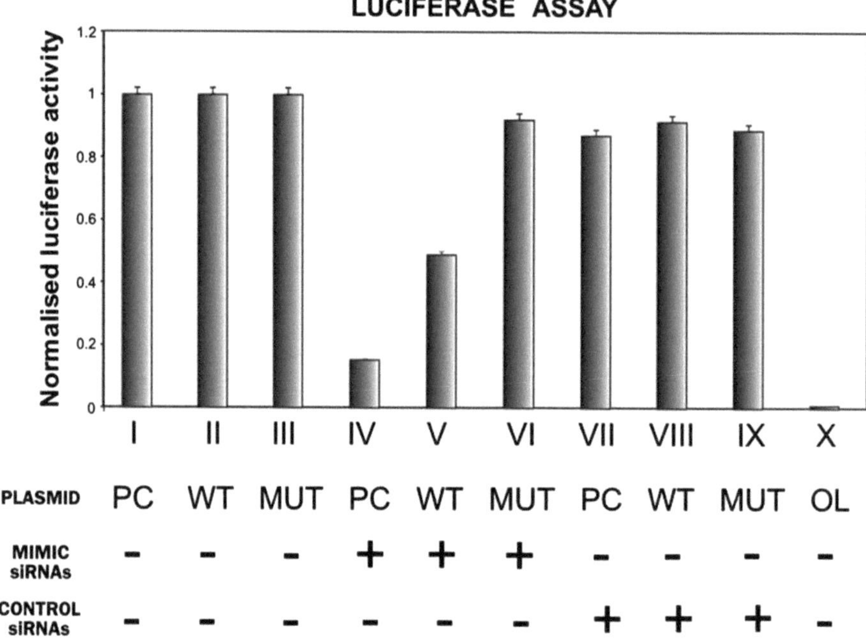

Fig. 4. Typical luminescence values in a positive miRNA target validation. *PC* positive control plasmid, *WT* wild-type plasmid, *MUT* mutant plasmid, and *OL* only lipofectamine.

8. To normalise the luminescence values, measure the total amount of protein using the BCA protein assay (Thermo Scientific). Prepare the working reagent by mixing 50 parts of BCA reagent A with one part of BCA reagent B (50:1, Reagent A:B). Then, transfer 10 μl of lysate of each sample to a 96-well plate, add 200 μl of the working solution and incubate for 30 min at 37°C. Finally, read the colorimetric reaction at 562 nm using the same multilabel counter.
9. In a positive scenario, when the candidate target sequence is a real target, a reduced luminescence in cells cotransfected with the wild-type plasmid and the mimicking siRNA is observed compared with both the wild-type plasmid with the control siRNA and the mutant plasmid cotransfected with both siRNAs. Relative reporter activity for mimic siRNA-treated cells is obtained by normalisation to non-siRNA-treated wild-type, mutant, or positive control constructs, respectively (Fig. 4).

4. Notes

1. The luciferase assay for miRNA target validation is based on the comparison of the luminescence production between the wild-type and mutant plasmids. Therefore, it is crucial that

the wild-type fragment amplified from genomic DNA contains exactly the same sequence as the original template, and the mutant plasmid only contains differences in the seed sequence target location compared with the wild-type. To avoid the introduction of an undesired mutation during PCR amplification, a high-quality polymerase enzyme, for instance Phusion polymerase, is recommended. However, one characteristic of Phusion polymerases is that they produce blunt ends, lacking the A-tailing activity present in other polymerases and preventing the cloning of PCR products in pGem-T. The pGem-T cloning system is based on this A-tailing activity, since the vector contains T-tail overhangs ready to pair with the A-tail added to the PCR products by regular Taq polymerases. A simple protocol for solving this problem consists of adding 1 μl of *Thermophilus aquaticus* polymerase (GoTaq, Promega) to the final products of the Phusion PCR and incubating for 15 min at 72°C. This Taq polymerase will add an A-tail to all the PCR products generated by the Phusion polymerase. In case that the target sequence contains introns, it must be amplified from cDNA.

2. The seed sequence of the miRNAs could be as little as a 7–8 mer site (34). It is recommended to mutate at least four of them. In this directed mutagenesis, the seed sequence can be modified to create a new restriction enzyme sequence, which afterwards facilitates the screening of positive colonies carrying the correct mutated fragment in the pGem-T easy plasmid.
3. The primers Reverse1 and Forward2 will create a small overlapping region between the two DNA fragments amplified from the first mutagenesis PCR that will allow the fusion of these fragments in a second PCR. The success of this second PCR will be determined by the length of each single fragment and the length of the overlapping region. So, for two DNA fragments of 200 bp, an overlapping region of at least 40–50 bp is necessary, and this can be generated by designing primers Reverse1 and Forward2 with 20 bp upstream and downstream of the mutation points (Fig. 2).
4. The selection of the cell line to perform the luciferase assay depends on two characteristics of the host cells. First, the cells have to be receptive to transfection using lipofectamine 2000 or any other transfection reagent that allows the simultaneous cotransfection of plasmid DNA and small RNAs. Second, performing the target validation assay using a cell line from the same species as the tested target sequence is recommended. This protocol has been successfully tested in NIH 3T3, Hela, and DF1 cell lines for mouse, human, and chicken target sequences, respectively.

5. The plasmid is usually purified from an overnight 100 ml *E. coli* culture using the appropriate size column (midiprep from the Qiagen purification kit). The transfection can also be carried out using plasmid DNA purified from smaller cultures using "miniprep kits"; however, this DNA has a lower transfection quality, which decreases the luciferase expression and increases the probability of contaminations in the cell culture.
6. An important issue affecting the reproducibility and accuracy of the luciferase values is the well being of the cell culture used for transfection. In this sense, it is important to use a nonstressed cell culture for the transfection. Always use a cell line culture that has been running for at least 3 weeks since the cells were initially plated from a frozen stock. The confluence of cells plated for transfection must always be the same to obtain reproducible results. An unnecessarily high cell confluence will produce a lower luciferase activity and a low confluence will produce an excessive variance in the luminescence values. Similarly, the amount of transfection reagent and the time that the cells are incubated with this reagent varies depending on the cell line selected for the transfection and has to be optimised before starting the assay. For instance, NIH 3T3 and Hela cells perform the assay well when they are incubated for 5–6 h with the lipofectamine mixture. However, the DF1 cell line becomes stressed when it is incubated for more than 4 h with the transfection reagent.
7. The final concentration of the mimicking siRNA has to be optimised before running the assay. The working concentration for siRNAs is usually between 10 and 100 nM. It can be optimised by performing a luciferase assay using only the positive control plasmid and different siRNA concentrations. The mimicking siRNA can be substituted for mimicking miRNA precursors. These precursors have a hairpin structure that mimics the original pre-miRNA generated by Drosha (see Subheading 1) and are incorporated into the silencing pathway at the Dicer cleavage step. The protocol for transfection and optimisation using these mimicking miRNA precursors is the same as the protocol described for siRNAs.
8. The homogenisation of the cell and lysis buffer can also be performed by scraping cells off the plate using a multichannel pipette and regular tips. Simply scrape homogenously the surface of the well with the tip for 1 min, aspirate the lysate up and down five times and then transfer it to a clean centrifuge tube.
9. The firefly luciferase (*Photinus pyralis*) catalyses an ATP-dependent oxidative decarboxylation of luciferin, producing a light emission at a wavelength of 562 nm. The kinetics of this luciferase-catalysed light emission presents a transient flash with peaks after 0.3–0.5 s and rapidly decays in a biphasic

manner. The reaction is finished within approximately 20 s, which requires rapid mixing and starting of the measurement of light emission. Measuring more than six wells in a row per reading using a manual luminometer is not recommended. This can be avoided when using an automated reader that injects the substrate into the well before the readings. It is also advisable to use a commercial substrate containing coenzyme A (Roche) that enhances light production by promoting the dissociation of oxidised luciferin, which is a potent inhibitor of the light reaction.

Acknowledgments

This work was supported by the Fundación Séneca (Murcia, Spain).

References

1. Bartel, D. P. (2004). MicroRNAs: genomics, biogenesis, mechanism, and function. *Cell* **116**, 281–297.
2. Bartel, B., and Bartel, D. P. (2003). MicroRNAs: at the root of plant development? *Plant Physiol* **132**, 709–717.
3. Basyuk, E., Suavet, F., Doglio, A., Bordonne, R., and Bertrand, E. (2003). Human let-7 stem–loop precursors harbor features of RNase III cleavage products. *Nucleic Acids Res* **31**, 6593–6597.
4. Baulcombe, D. (2004). RNA silencing in plants. *Nature* **431**, 356–363.
5. Kloosterman, W. P., and Plasterk, R. H. (2006). The diverse functions of microRNAs in animal development and disease. *Dev Cell* **11**, 441–450.
6. Lee, Y., Jeon, K., Lee, J. T., Kim, S., and Kim, V. N. (2002). MicroRNA maturation: stepwise processing and subcellular localization. *EMBO J* **21**, 4663–4670.
7. Zeng, Y., and Cullen, B. R. (2003). Sequence requirements for microRNA processing and function in human cells. *RNA* **9**, 112–123.
8. Lund, E., Guttinger, S., Calado, A., Dahlberg, J. E., and Kutay, U. (2004). Nuclear export of microRNA precursors. *Science* **303**, 95–98.
9. Yi, R., Qin, Y., Macara, I. G., and Cullen, B. R. (2003). Exportin-5 mediates the nuclear export of pre-microRNAs and short hairpin RNAs. *Genes Dev* **17**, 3011–3016.
10. Grishok, A., and Sharp, P. A. (2005). Negative regulation of nuclear divisions in *Caenorhabditis elegans* by retinoblastoma and RNA interference-related genes. *Proc Natl Acad Sci U S A* **102**, 17360–17365.
11. Hutvagner, G., and Zamore, P. D. (2002). A microRNA in a multiple-turnover RNAi enzyme complex. *Science* **297**, 2056–2060.
12. Leuschner, P. J., Ameres, S. L., Kueng, S., and Martinez, J. (2006). Cleavage of the siRNA passenger strand during RISC assembly in human cells. *EMBO Rep* **7**, 314–320.
13. Matranga, C., Tomari, Y., Shin, C., Bartel, D. P., and Zamore, P. D. (2005). Passenger-strand cleavage facilitates assembly of siRNA into Ago2-containing RNAi enzyme complexes. *Cell* **123**, 607–620.
14. Rand, T. A., Petersen, S., Du, F., and Wang, X. (2005). Argonaute2 cleaves the anti-guide strand of siRNA during RISC activation. *Cell* **123**, 621–629.
15. Bagga, S., et al. (2005). Regulation by let-7 and lin-4 miRNAs results in target mRNA degradation. *Cell* **122**, 553–563.
16. Eulalio, A., Huntzinger, E., and Izaurralde, E. (2008). GW182 interaction with Argonaute is essential for miRNA-mediated translational repression and mRNA decay. *Nat Struct Mol Biol* **15**, 346–353.
17. Parker, J. S., Roe, S. M., and Barford, D. (2006). Molecular mechanism of target RNA transcript recognition by Argonaute-guide complexes. *Cold Spring Harb Symp Quant Biol* **71**, 45–50.
18. Petersen, C. P., Bordeleau, M. E., Pelletier, J., and Sharp, P. A. (2006). Short RNAs repress

translation after initiation in mammalian cells. *Mol Cell* **21**, 533–542.

19. Pillai, R. S., et al. (2005). Inhibition of translational initiation by Let-7 MicroRNA in human cells. *Science* **309**, 1573–1576.
20. Rhoades, M. W., Reinhart, B. J., Lim, L. P., Burge, C. B., Bartel, B., and Bartel, D. P. (2002). Prediction of plant microRNA targets. *Cell* **110**, 513–520.
21. Davis, E., et al. (2005). RNAi-mediated allelic trans-interaction at the imprinted Rtl1/Peg11 locus. *Curr Biol* **15**, 743–749.
22. Yekta, S., Shih, I. H., and Bartel, D. P. (2004). MicroRNA-directed cleavage of HOXB8 mRNA. *Science* **304**, 594–596.
23. Lim, L. P., Glasner, M. E., Yekta, S., Burge, C. B., and Bartel, D. P. (2003). Vertebrate microRNA genes. *Science* **299**, 1540.
24. Baek, D., Villen, J., Shin, C., Camargo, F. D., Gygi, S. P., and Bartel, D. P. (2008). The impact of microRNAs on protein output. *Nature* **455**, 64–71.
25. Sethupathy, P., Megraw, M., and Hatzigeorgiou, A. G. (2006). A guide through present computational approaches for the identification of mammalian microRNA targets. *Nat Methods* **3**, 881–886.
26. Lewis, B. P., Shih, I. H., Jones-Rhoades, M. W., Bartel, D. P., and Burge, C. B. (2003). Prediction of mammalian microRNA targets. *Cell* **115**, 787–798.
27. Lim, L. P., et al. (2005). Microarray analysis shows that some microRNAs downregulate large numbers of target mRNAs. *Nature* **433**, 769–773.
28. Elmen, J., et al. (2008). Antagonism of microRNA-122 in mice by systemically administered LNA-antimiR leads to up-regulation of a large set of predicted target mRNAs in the liver. *Nucleic Acids Res* **36**, 1153–1162.
29. Nicolas, F. E., et al. (2008). Experimental identification of microRNA-140 targets by silencing and overexpressing miR-140. *RNA* **14**, 2513–2520.
30. Beitzinger, M., Peters, L., Zhu, J. Y., Kremmer, E., and Meister, G. (2007). Identification of human microRNA targets from isolated argonaute protein complexes. *RNA Biol* **4**, 76–84.
31. Karginov, F. V., et al. (2007). A biochemical approach to identifying microRNA targets. *Proc Natl Acad Sci USA* **104**, 19291–19296.
32. Selbach, M., Schwanhausser, B., Thierfelder, N., Fang, Z., Khanin, R., and Rajewsky, N. (2008). Widespread changes in protein synthesis induced by microRNAs. *Nature* **455**, 58–63.
33. Vinther, J., Hedegaard, M. M., Gardner, P. P., Andersen, J. S., and Arctander, P. (2006). Identification of miRNA targets with stable isotope labeling by amino acids in cell culture. *Nucleic Acids Res* **34**, e107.
34. Brennecke, J., Stark, A., Russell, R. B., and Cohen, S. M. (2005). Principles of microRNA-target recognition. *PLoS Biol* **3**, e85.

Chapter 12

Experimental Identification of MicroRNA Targets by Immunoprecipitation of Argonaute Protein Complexes

Michaela Beitzinger and Gunter Meister

Abstract

MicroRNAs (miRNAs) represent a class of small noncoding RNAs that negatively regulate gene expression. Intensive research during the past decade has established miRNAs as key regulators of many cellular pathways. MiRNAs have also been implicated in a number of diseases including various forms of cancer. Mammalian miRNAs associate with members of the Argonaute (Ago) protein family and function in multiprotein complexes. MiRNAs guide Ago protein complexes to partially complementary sequences typically located in the 3′ untranslated region (UTR) of their target mRNAs leading to the inhibition of its translation and/or its destabilization. To understand the biological roles of miRNAs, it is essential to identify the mRNA targets that they regulate. Because of the low degree of complementarity between the miRNA and its target sequence, it is often difficult to find targets computationally. Therefore, biochemical methods are needed to identify miRNA targets experimentally. The availability of highly specific monoclonal antibodies against Argonaute proteins allows for the isolation of functional Ago-miRNA–mRNA complexes from different cell lines, tissues, or even patient samples. Here we provide a detailed protocol for isolation and identification of miRNA target mRNAs from immunoprecipitated human Ago protein complexes.

Key words: Argonaute proteins, MicroRNAs, Immunoprecipitation, Gene silencing, mRNAs, Dicer, Drosha

1. Introduction

miRNAs are conserved from plants to human and act as powerful regulators of gene expression. During the last decade, miRNAs have been studied in detail and it has been found that they function in various biological processes including organ development, cell differentiation, cell cycle, and apoptosis (1–4). miRNA genes are transcribed by RNA polymerases II or III generating primary miRNA transcripts that are processed by the RNase III enzyme Drosha as part of the nuclear microprocessor complex. Drosha produces stem-loop structured miRNA precursors (pre-miRNAs) that are

Tamas Dalmay (ed.), *MicroRNAs in Development: Methods and Protocols*, Methods in Molecular Biology, vol. 732, DOI 10.1007/978-1-61779-083-6_12, © Springer Science+Business Media, LLC 2011

subsequently transported to the cytoplasm by the export receptor Exportin 5. In the cytoplasm, another RNase III enzyme, termed Dicer, cleaves off the loop of the pre-miRNA resulting in a short-lived double-stranded (ds) miRNA/miRNA* RNA intermediate. Such short dsRNAs are unwound and only one strand is incorporated into miRNA–protein complexes [often referred to as miRNPs or miRNA-containing RNA-induced silencing complex (miRISC)] and gives rise to the mature miRNA. The miRNA* strand, however, is degraded by cellular nucleases (Fig. 1) (1, 4).

MiRNAs recognize partially complementary binding sites located in the 3′ UTR of target mRNAs and it has been shown that miRNPs can regulate gene expression in different ways. On target sites with a high degree of complementarity, miRNAs can induce sequence-specific cleavage of the target mRNA. Targets with low degree of complementarity can either be destabilized by recruiting deadenylation and decapping enzymes or their translation is repressed without altering mRNA levels (2).

Members of the Ago protein family are at the core of miRNPs and represent the actual mediators of gene silencing, whereas miRNAs are viewed as guides that direct Ago proteins to distinct

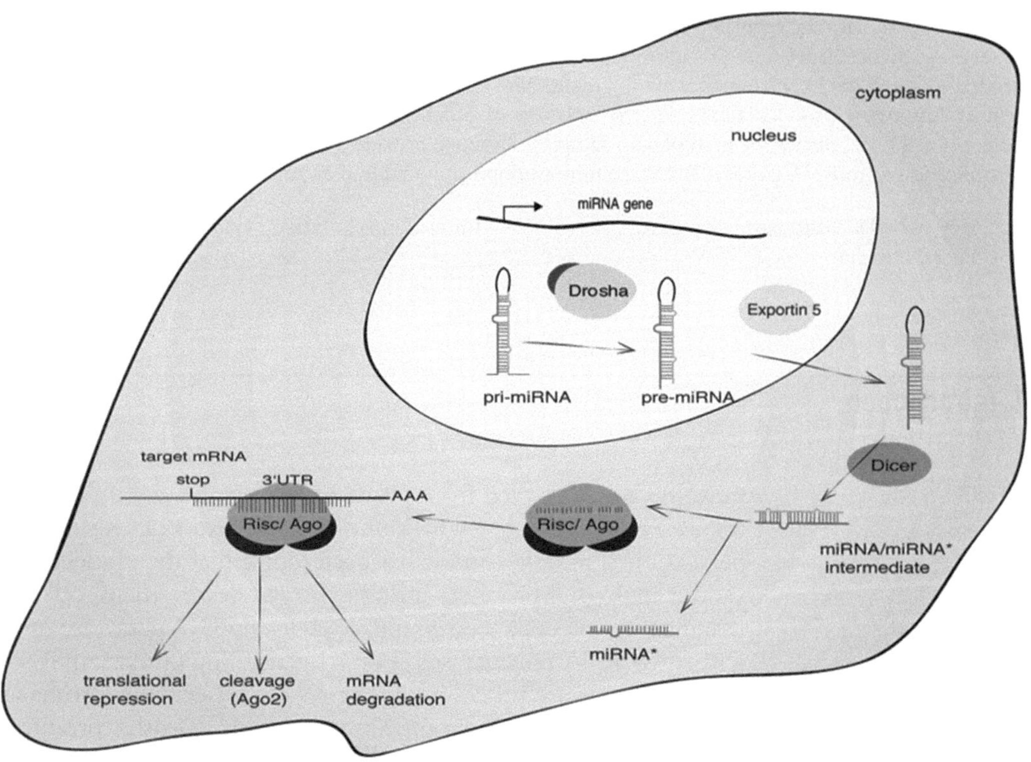

Fig. 1. Schematic overview over the miRNA pathway. MiRNA genes are transcribed as primary transcripts, which are processed by the RNase III enzymes Drosha and Dicer to mature, single-stranded miRNAs. MiRNAs bind members of the Ago protein family to form the RNA-induced silencing complex. The miRNA guides RISC to partially complementary target sequences located in the 3′ UTR of target mRNAs, which leads to mRNA destabilization and/or translational repression.

target RNAs. Ago proteins are highly conserved and contain PAZ (PIWI–Argonaute–zwille) and PIWI (P-element-induced wimpy testes) domains. Numerous structural studies on archaeal and bacterial Ago proteins revealed that the PAZ domain forms a highly specific binding module for the 3′ end of small RNAs. In contrast, the characteristic folding of PIWI domains is similar to RNase H and can function as endonucleases. Indeed, it has been demonstrated that some Ago proteins are endonucleases and have, therefore, been termed "Slicer." In mammals, only Ago2 seems to have endonuclease activity, although critical amino acids are conserved in other Ago proteins as well (5–8).

Due to the low degree of complementarity between miRNAs and target mRNAs, it is difficult to identify miRNA target mRNAs (1, 9). A common strategy to find target mRNAs is based on computational predictions. It has been found that six to seven nucleotides at the 5′ end of the miRNA are the most important determinants for target specificity. Such sequences are called "seed sequence." Based on seed sequence conservation and free binding energy, miRNA targets have been predicted in a variety of different organisms. The different prediction programs predict a vast number of different targets for one given miRNA. Some algorithms even predict one-third of all human mRNAs as miRNA targets. Computer programs usually predict targets on a genome-wide scale irrespective of tissue-specific mRNA and miRNA expression patterns. Therefore, it is very difficult to find miRNA targets that are specific to tissues or cell lines. It is becoming more and more apparent that it is very important to develop biochemical tools based on physical interactions between miRNPs and their specific target mRNAs to reliably identify subsets of miRNA targets.

To date, several groups reported the identification of miRNA targets by co-immunoprecipitation of Ago proteins together with associated mRNAs (10–14). In addition to that, co-immunoprecipitation of miRNA targets bound to Ago proteins is also an effective tool for the identification or validation of miRNA target candidates (15).

This protocol describes and illustrates the identification of miRNA targets using co-immunoprecipitation with Ago proteins and further analysis of targets by gene expression array technology and/or qRT-PCR.

2. Materials

2.1. Cell Culture and Lysis

1. Dulbecco's Modified Eagle's Medium (DMEM) (PAA) supplemented with 10% fetal bovine serum (Biochrome) and 1× penicillin–streptomycin (PAA).
2. Phosphate-buffered saline pH 7.5 (PBS).

3. Cell lysis buffer: 0.5% NP40, 150 mM KCl, 25 mM Tris–glycine pH 7.5, 2 mM EDTA, and 0.5 mM DTT (add immediately before use).
4. Cell scrapers (Roth).

2.2. Antibodies and Immunoprecipitation

1. Anti-Ago2 (11A9), anti-Ago1 (4B8) hybridoma supernatant (10, 16) [antibodies available from http://www.ascenion.de, appropriate rat monoclonal control antibodies of the same subtype (IgG-2a)] (hybridoma supernatant).
2. Protein-G-sepharose beads (GE-Healthcare).
3. IP-wash buffer: 300 mM NaCl, 50 mM Tris–glycine pH 7.5, 5 mM $MgCl_2$, 0.05% NP40.

2.3. Detection of Ago2 from Anti-Ago2 Immunoprecipitations

1. Anti-Ago2 (11A9) hybridoma supernatant (antibodies available from http://www.ascenion.de), anti-Rat-HRP antibody (GE-Healthcare).
2. TBS: 150 mM NaCl, 10 mM Tris–glycine pH 8.0.
3. Hybond ECL membrane (GE-Healthcare).

2.4. Isolation of mRNAs from Anti-Ago Immuoprecipitations

1. Proteinase K digestion buffer: 300 mM NaCl, 200 mM Tris–glycine pH 7.5, 25 mM EDTA, 2% SDS.
2. Proteinase K (Applichem).
3. Phenol/chloroform/isoamyl alcohol for RNA extraction (Roth).
4. Phenol (Roth).
5. Glycogen (Fermentas).

2.5. Northern Blot for Detection of Co-immuno-precipitated miRNAs

1. EDC-crosslinking solution: 0.16 M EDC prepared in 0.13 M 1-methimidazole pH 8.0.
2. G25 spin column (GE-Healthcare).
3. γ-^{32}P-ATP (Perkin–Elmer).
4. T4 poly-nucleotide kinase (NEB).
5. Hybond-N+ nylon membrane (GE-Healthcare).
6. 20× SSC pH 7.0 (3 M NaCl and 0.3 M sodium citrate).
7. 50× Denhardt's solution: 1 % Albumin Fraktion V, 1% Polyvinyl-pyrrolidon K30, and 1% Ficoll 400.
8. Prehybridization solution: 7.5 ml 20× SSC, 600 μl 1 M Na_2HPO_4 pH 7.2, 21 ml 10% SDS, 300 μl 50× Denhardt's solution.
9. Wash buffer 1: 1× SSC, 5% SDS.
10. Wash buffer 2: 1× SSC, 1% SDS.

2.6. cDNA Synthesis, Gene Expression Array and qPCR

1. DNAse I (Fermentas).
2. Ribolock (Fermentas).
3. First-strand cDNA synthesis Kit (Fermentas).
4. Eurogentec 2× Mesagreen qPCR Mix (with fluorescein for Biorad iCycler).
5. Affymetrix human gene expression array U133 Plus and reagents.

3. Methods

For efficient immunoprecipitation, the use of fresh cells is preferred, but if necessary cells could also be stored. Snap freeze the dry cell pellet in liquid nitrogen and store it at –80°C until use. For gene expression array analysis use 5–10 mg of total protein (see Notes 1–3).

3.1. Beads Preparation and Antibody Coupling

1. Before use, the beads need to be washed once with PBS (see Note 4). Pipet 80 μl protein-G-sepharose beads (50% slurry) in a 1.5-ml tube, add 1 ml PBS, and centrifuge at 1,000×*g* for 1 min at 4°C. Remove and discard supernatant. The supernatant should be taken off carefully to avoid loss of beads.
2. Use 1 ml hybridoma supernatant of Ago2 (11A9), Agol (4B8) antibody or unspecific control antibody (e.g. anti-FLAG), per 80 μl protein-G-sepharose beads (50% slurry), and incubate at 4°C under rotation over night or at least for 2 h (see Note 5).
3. After incubation with the antibody, centrifuge at 1,000×*g* for 1 min at 4°C and remove the supernatant. Wash the beads once with IP wash buffer and resuspend in 500 μl PBS. The sepharose pellet is sometimes difficult to see and therefore, it is better to leave a little liquid (some μl) in the tube to avoid loss of beads.

3.2. Sample Preparation

1. If adherent cells (e.g., HEK293 or Hela) are used, one 15-cm plate will typically result in 1 mg of total protein. Each 15-cm plate should be washed with 5 ml PBS to remove residual culture medium. To remove all residual PBS, the plates can be sloped to summon residual liquid at the edge of the plate. Carefully remove all remaining liquid from the plate.
2. For each 15-cm plate use 1 ml of cold (4°C) lysis buffer. The lysis can be performed directly on the plates. Ensure that the lysis buffer covers the whole cell layer and store the plates at 4°C for 30 min.

3. Alternatively, cells can be scraped off the plate in 5 ml PBS, centrifuge them at 200 × *g* and 4°C for 10 min. Cells can then be lysed directly in the centrifugation tube. Add 1 ml lysis buffer, resuspend and incubate for 30 min on ice. The protocol can be adjusted easily to suspension culture cells. Simply centrifuge the cells, wash once with PBS and lyse the pellet in the appropriate volume of lysis buffer.
4. After lysis, centrifuge at 17,000 × *g* (maximum speed in a tabletop centrifuge) at 4°C for 30 min. Transfer the supernatant into a new tube. Make sure that the lysate is clear, if not centrifuge again to remove any residual cell debris (see Note 6).
5. For the immunoprecipitation, the lysate should be aliquoted to equal volumes per antibody used (e.g., Ago2 and ctrl. antibody). For qRT-PCR analysis, 10% of the lysate volume, which is used for the immunoprecipitation, should be saved as input sample. This is important for data analysis as well as quality control.
6. The input sample should be stored on ice until the immunoprecipitation is ready for RNA precipitation.
7. Add beads prepared as described in Subheading 3.1 to each immunoprecipitation reaction and fill the tube with cell lysis buffer. Avoid larger volumes of air in the tube and incubate about 3 h at 4°C under rotation (see Note 7).
8. After incubation with cell lysate, the beads are washed gently with IP-wash buffer. Therefore, pellet the beads at 1,000 × *g* for 1 min at 4°C. Carefully remove as much of the supernatant as possible and avoid soaking up beads into the pipette tip. If the immunoprecipitation is performed in round-bottom tubes (e.g., 2-ml reaction tubes), transfer the first wash into conical tubes for a better visibility of the protein-G-beads.
9. Add 1 ml IP-wash buffer and centrifuge again. Repeat the procedure at least five times.
10. Perform one additional washing step with PBS to remove residual detergent and transfer beads in a new tube to remove protein and/or RNA contamination that might bind to the walls of the tube (see Note 8).
11. A small aliquot (~100 μl) of beads in PBS (last washing step in ~1 ml total volume) should be saved and transferred to a new tube for western blot and/or northern blot analysis to confirm efficient precipitation of Ago protein complexes and co-immunoprecipitated miRNAs. The beads should be centrifuged at 1,000 × *g* for 1 min at 4°C and resuspended in 2× SDS-loading dye (western blot). For miRNA detection (northern blot) follow the protocol described in subheading 3.4 and 3.6. The final RNA pellet can be solved in 2x RNA loading dye instead of RNAse free H_2O. The samples can be stored at −20°C until use.

3.3. Western Blot Analysis of Anti-Ago2 Immunoprecipitations

1. For the detection of Ago proteins in the immunoprecipitates, follow standard western blot protocols with some specifications.
2. The stored lysates in SDS-loading dye should be thawed on ice and then incubated for 5 min at 95°C. Load the samples on a 10% SDS–PAGE and include one well for a prestained molecular weight marker. Pipet protein samples from the top, because beads are still present in the sample. Avoid loading of beads onto the gel, which leads to poor protein separation.
3. For protein detection, monoclonal rat-anti-Ago2 (11A9) hybridoma supernatants are diluted 1:50 (e.g., TBST 0.1% Tween with 5% dry milk powder).
4. Transfer proteins separated by SDS–PAGE onto nitrocellulose membrane using semidry electroblotting.
5. Incubate the membrane for 1 h at room temperature or overnight at 4°C.
6. Wash the membrane and incubate with an anti-rat-HRP antibody for 1 h at room temperature.
7. Follow the standard ECL protocol to detect the immunoprecipitated Ago protein signals on the western blot membrane (an example is shown in Fig. 2a).

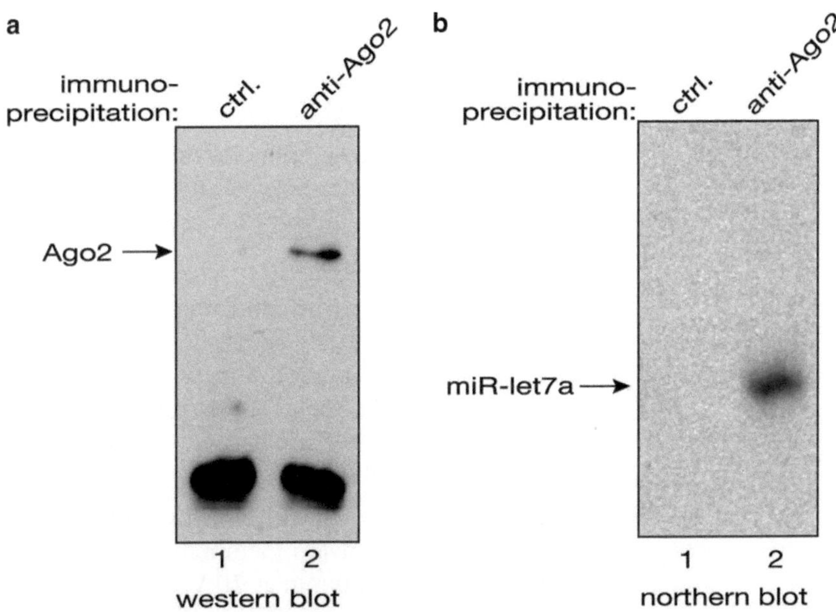

Fig. 2. Validation of anti-Ago2 immunoprecipitation efficiency. Endogenous Ago2 was immunoprecipitated from 2 mg HEK 293T (**a**) or HeLa (**b**) cell lysate using monoclonal anti-Ago2 (11A9) antibodies. Unrelated antibodies (anti-FLAG) served as controls. Immunoprecipitated endogenous Ago2 was analyzed by western blotting analysis (**a**). Co-immunoprecipitation of Ago2-associated miRNAs was confirmed by northern blot analysis (**b**).

3.4. RNA Precipitation

1. After the last wash, the remaining protein-G-sepharose are pelleted at 1,000 × *g* for 1 min at 4°C and resuspended in 250 μl Proteinase K buffer supplemented with 10 μl proteinase K (20 mg/ml). Input samples are treated equivalently. Incubate samples at 65°C for 15 min with gentle agitation to keep beads in motion.
2. 300 μl phenol (for RNA-extraction) is added and the samples are mixed by vortexing. To separate the phenol from the aqueous phase, centrifuge at 17,000 × *g* (maximum speed) for 10 min at 4°C (see Note 9).
3. The upper aqueous phase is transferred to a new 1.5-ml tube. Repeat the extraction once with chloroform/isoamyl alcohol to remove residual traces of phenol, which might interfere with subsequent RNA analysis. For final RNA precipitation, add 1 μg 20 mg/ml glycogen and 2.5× vol. 100% ethanol, mix well and store at −20°C over night (do not add additional salt for the precipitation).

3.5. Northern Blot for miRNA Detection

For the detection of co-immunoprecipitated miRNAs, use standard northern blot protocols and EDC cross linking (17). In brief:

1. Before starting the northern blot, prepare a radioactively labeled probe for a miRNA known to be expressed in the cells used (e.g. let-7a in Hela cells). Use 1 μl of 20 μM oligonucleotide (sequence is antisense to the miRNA, which is analyzed) and 1 μl γ-^{32}P-ATP (3,000 Ci/mmol) in a 20 μl T4-PNK reaction and incubate at 37°C for 30 min. Add 30 μl 30 mM EDTA and purify with G25 spin column (see Note 10).
2. For the detection of co-immunoprecipitated miRNAs, pour a 12% denaturing polyacrylamide gel and let it set for at least 2 h.
3. Prepare the gel run in 1× MOPS buffer and pre-run the gel for at least 10 min at 200 V.
4. Thaw stored samples on ice and boil 5 min at 95°C. Rinse wells of the gel immediately before use to remove urea, which diffuses from the gel and accumulates in the wells of the gel (see Note 11). Load samples and run the gel at 200–400 V. Mature miRNAs typically run between the bromophenol blue and the xylene cyanol dye front.
5. Prepare the blot sandwich with 3MM Whatman papers (three on each site), gel and nylon membrane in a semidry blotting chamber and run for 30 min at 20 V.
6. While blotting, prepare prehybridization solution and prewarm to 50°C.
7. Immediately before use, prepare the EDC cross-linking solution. Saturate a sheet of 3MM Whatman paper (approximately

the size of the gel) and place the gel (RNA side up) on top, wrap in saran wrap, and incubate for 1 h at 50°C. After cross-linking, wash the membrane in distilled water for a few seconds.

8. Place the membrane in a hybridization bottle and incubate in prehybridization solution for 1 h at 50°C in a hybridization oven under rotation.
9. Add the γ-^{32}P-ATP-labeled probe directly into the prehybridization solution and incubate over night at 50°C under rotation.
10. Wash the membrane two times in wash buffer 1 and once in wash buffer 2 for 10 min each at 50°C.
11. Wrap the membrane with saran wrap and expose to X-ray film for several hours or days at –80°C. An example for the detection of Ago2-associated miRNAs is shown in Fig. 12.2b.

3.6. mRNA Analysis

Centrifuge at 17,000×*g* (maximum speed) for 30 min at 4°C to pellet RNA. Air dry the pellet and dissolve in RNAse-free H_2O or directly in DNAse I digestion solution depending on further analysis. Use qRT-PCR to analyze specific mRNAs. Gene expression array analysis should be performed to identify unknown targets that copurify with anti-Ago antibodies.

3.6.1. qRT-PCR Analysis

1. DNAse I digestion

   Dissolve the RNA-pellet in 10 µl DNAse I digestion reaction solution (Fermentas) use 1 µl DNAse I, 1 µl DNAse I buffer, and 0.25 µl Ribolock and add H_2O to a final volume of 10 µl. Incubate for 30 min at 37°C. Add 1 µl of 25 mM EDTA and incubate for 10 min 65°C to inactivate DNAse I.

2. cDNA Synthesis

   Set up a 20 µl reaction according to the suppliers protocol (Fermentas cDNA Synthesis Kit) and incubate for 1 h at 37°C. The resulting 20 µl of cDNA could be diluted to a final volume of 50 µl H_2O.

3. qPCR

   For a typical qPCR reaction, 1 µl of diluted cDNA per well is sufficient. To minimize the measuring of pipetting errors, it is recommended to measure in triplicates. The easiest way for this is to prepare a threefold submix, which is aliquoted into three wells of a PCR plate.

Unspecific binding of RNA to the protein-G-sepharose beads cannot be absolutely avoided and the amount of specific mRNA, which is immunoprecipitated is often limited especially for low abundant target mRNAs. Therefore, we are calculating the enrichment of a known or potential Ago target relative to GAPDH or t-RNA enrichment, since those seem not to be targets by Ago

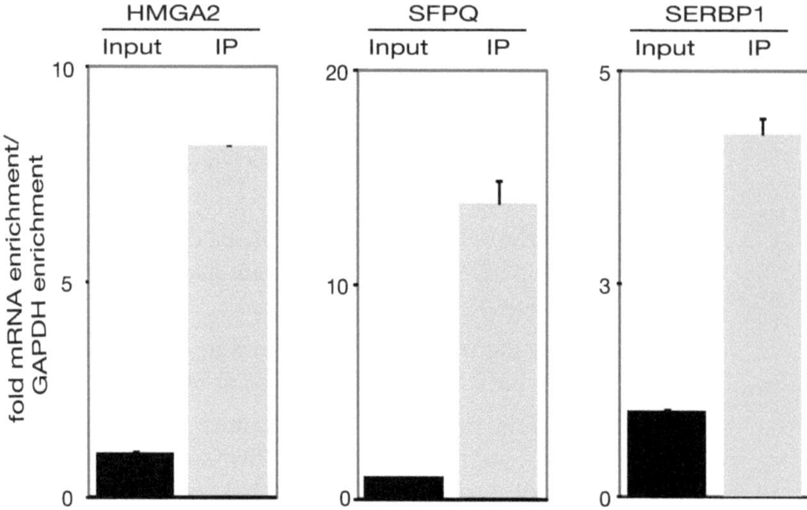

Fig. 3. qRT-PCR analysis of Ago2-associated mRNAs. Endogenous Ago2 was immunoprecipitated from 2 mg total HeLa cell lysate with monoclonal anti-Ago2 antibodies. An unrelated control antibody (anti-FLAG) served as control. The co-immunoprecipitation of Ago2-associated mRNAs was confirmed by qRT-PCR of the known miRNA targets HMGA2, SERBP1, and SFPQ. The fold enrichment was calculated in relation to the amount of input RNA and to the unspecific enrichment of GAPDH mRNA levels.

protein in the cell lines analyzed here. However, other mRNAs that are not under the control of the miRNA pathway could be used for normalization as well (an example for the detection of known miRNA targets is shown in Fig. 3).

3.6.2. Gene Expression Array Analysis

Gene expression array analysis depends on many factors such as the sort of array used (U133, Gene-ST, etc.) or the system (Affymetrix, Agilent, etc.). Since all standard array systems are useful here, the protocol does not go into detail and describes in brief the data analysis resulting from the use of gene expression array U133 Plus (Affymetrix).

Precipitated RNA was labeled and processed with the Gene Chip (Affymetrix) kit according to the manufacturer's instructions. Samples were hybridized to Human Genome U133 Plus 2.0 arrays.

Microarray data were analyzed with Agilent Genespring software. Expression values below 0.01 were set to 0.01. Each measurement was divided by the 50th percentile of all measurements in that sample. All immunoprecipitated samples were normalized to the corresponding control immunoprecipitates. Each measurement for each gene in the immunoprecipitated samples was divided by the median of that gene's measurements in the corresponding control immunoprecipitates. Using this normalization procedure, the normalized expression value of each transcript in anti-Ago immunoprecipitates reflects its over- or underrepresentation in the immunoprecipitated transcript pool relative to unspecific binding of mRNAs to sepharose beads. Further data analysis was performed by filtering anti-Ago immunoprecipitated samples with raw measurements over 100 that were 20-fold enriched (Fig. 4) or 40-fold (Table 1) in anti-Ago immunoprecipitates compared with control immunoprecipitations.

4. Notes

1. It is critical to use RNA-free reagents since RNase contamination affects the RNA quality and is very often the reason for irreproducible data.
2. The amount of cells needed varies depending on the preferred analysis. If only qRT-PCR is performed, the amount of lysate can be reduced to 1–2 mg of total protein. The amount of cells needed differs because gene expression array analysis does not include amplification as PCR does. Therefore, less amount of lysate is sufficient for performing qPCR analysis.
3. Be sure that the cells become not too confluent to avoid cellular stress. Changing conditions in the cell growth could result in stress-mediated gene regulation, which could also affect Ago mRNA association and, therefore, the reproducibility of the results.
4. For immunoprecipitation, the protein-G-sepharose beads need to be prepared and coupled with the antibody. It is advisable to use pipette tips with the end cut off to prevent damage to the beads. Therefore, cut off about 5 mm of the tips. For each immunoprecipitation use 40 μl packed beads.
5. For qRT-PCR analysis, the amount of beads and antibody used could be scaled down or up depending on the scale of the experiment.
6. Make sure that the lysate is clear, if not centrifuge again to get rid of any residual cell debris. This is a critical point for the success of the experiment.
7. Avoid larger volumes of air in the tube, because proteins start to aggregate preferentially at the air–liquid interphase.
8. Transfer beads after the final washing step into fresh tubes to remove protein and/or RNA contaminations that bind to the walls of the tube.
9. If pure phenol is used for RNA extraction, beads will be visible as a white pellet after centrifugation. If a phenol/chloroform mixture is used for extraction, beads will be found in the interphase, which makes the transfer of the aqueous phase more difficult.

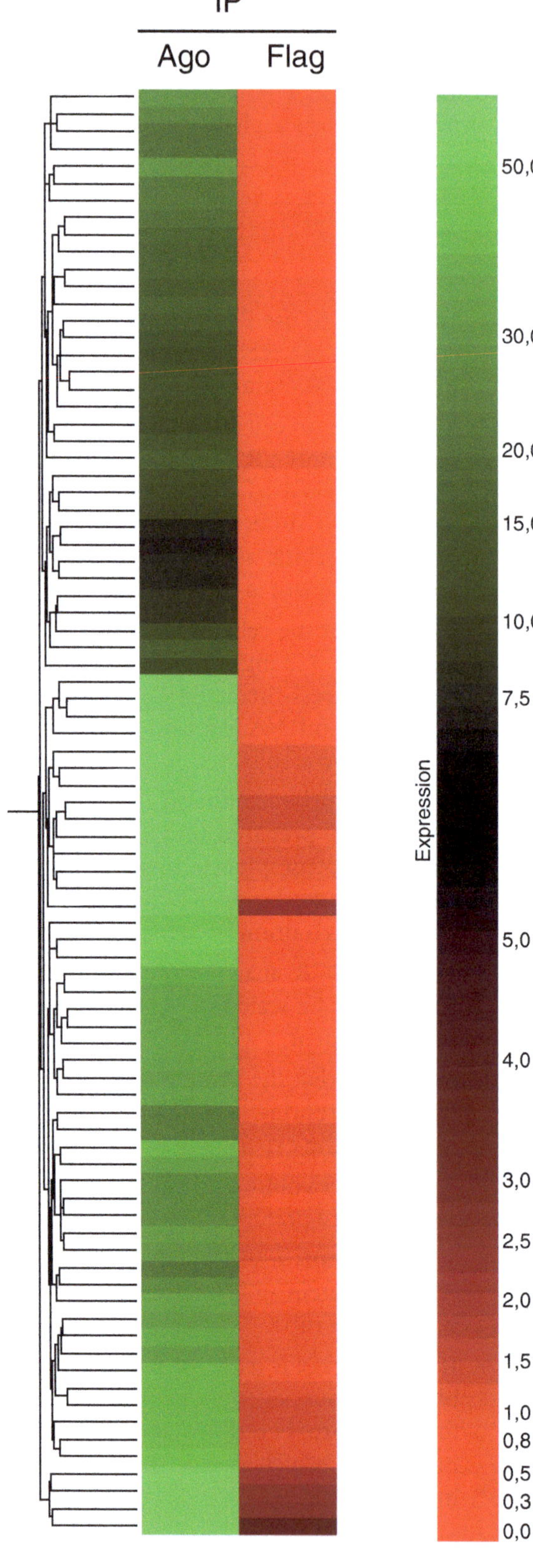
IP
Ago
Flag
50,0
30,0
20,0
15,0
10,0
7,5
Expression
5,0
4,0
3,0
2,5
2,0
1,5
1,0
0,8
0,5
0,3
0,0

Table 1
Ago1-associated mRNAs isolated from the B-cell line BL41

Fold enrichment	Prediction algorithms				Targets	
Ago1 IP vs. FLAG IP	miRanda	miRBase	Targetscan	Pictar	Gene	Accession no.
175.0	+	+	Poor	–	KIAA1641	AA632139
164.8	+	+	+	+	BACH2	NM_021813
152.2	–	–	–	–		AW274846
151.1	+	+	+	–	ZNF6	AU157017
130.9	+	+	Poor	–	KLRC2	NM_002260
106.8	–	–	–	–		N66584
98.3	–	–	–	–	MGC27005	U31738
90.8	+	+	Poor	–	RGS13	AF030107
89.0	+	+	Poor	–	KLRC3	NM_002261
88.4	+	+	Poor	–	MGST1	AI220117
85.0	+	+	+	+	BCL6	NM_001706
82.4	+	+	+	+	GPM6A	BF939489
80.2	+	+	+	+	OSBPL10	NM_017784
69.2	+	+	+	+	ZNF804A	AF052145
69.2	+	+	+	+	JAZF1	AL047908
69.0	+	+	+	+	MGC8685	AL533838
67.7	+	+	+	–	KIAA0774	AI818409
66.2	+	+	+	+	RBMS1	AL517946
66.2	–	–	–	–		AK000776
64.9	–	–	–	–		W73921
64.3	–	–	Poor	–	BCNP1	BE646461
64.2	+	–	–	+	TM4SF13	NM_014399
64.0	+	+	+	+	MME	AI433463

(continued)

Fig. 4. Gene expression array analysis of Ago1-associated mRNAs isolated from BL41 cells. Endogenous Ago1 was immunoprecipitated from 10 mg BL41 total cell lysate with a monoclonal anti-Ago1 (4B8) antibody. An unrelated anti-FLAG antibody served as control. Associated mRNAs were isolated and processed for Affymetrix U133 plus array. Gene lists were generated with restrictions to raw value of anti-Ago1 immunoprecipitation more than 100 and a minimal 20-fold enrichment compared with the control immunoprecipitation. *Green*: high-signal intensity, *red*: low-signal intensity.

Table 1 (continued)

Fold enrichment	Prediction algorithms				Targets	
Ago1 IP vs. FLAG IP	miRanda	miRBase	Targetscan	Pictar	Gene	Accession no.
63.8	+	+	Poor	–	NRN1	NM_016588
63.3	+	–	+	+	CNR1	U73304
63.1	+	+	+	+	IBRDC2	AI953847
62.2	+	–	+	+	HDGFRP3	AK001280
58.5	–	–	–	–		R49295
56.4	–	+	Poor	–	RNF175	AW051591
55.8	+	+	+	+	ZNF533	AI694320
55.5	+	+	+	+	SLC25A27	AW779950
53.3	–	+	–	–	C8orf12	NM_054017
53.0	+	+	+	+	HIPK2	BF218115
49.6	–	+	+	+	COCH	AA669336
49.2	–	–	Poor	–	OSAP	AF329088
48.0	+	+	+	+	PCDH9	AI524125
47.3	–	–	+	+	SGEF	AI989530
47.0	–	+	+	+	STK39	NM_013233
45.2	–	+	+	+	DPYD	NM_000110
44.8	+	+	+	+	ID4	AW157094
40.7	+	+	Poor	+	CTPS2	AK025654
40.5	+	–	+	+	FLJ20366	NM_017786

Endogenous Ago1 was immunoprecipitated from 10 mg BL41 total cell lysate with monoclonal anti-Ago1 (4B8) antibody. An unrelated anti-FLAG antibody served as control. Associated mRNAs were isolated and hybridized to Affymetrix U133 plus array. Gene lists were generated with restrictions to raw values of Ago2 immunoprecipitation greater than 100 and to a minimal 40-fold enrichment compared with the control immunoprecipitation. These restrictions resulted in a list of 42 genes where only six (*gray*) are not predicted by any of the prediction programs pictar, targetscan, miranda, or miRbase. Fold enrichment depicts the enrichment of signal intensity in anti-Ago2 immunoprecipitations compared with control immunoprecipitations (poor = predicted only for poorly conserved sites)

10. The radioactive-labeled probe for northern blot analysis can be stored at –20°C until use but no longer than a few days, since ^{32}P has only a half-life of 14 days.
11. Rinse the wells of the denaturing polyacrylamide gel immediately before loading the samples. Urea usually diffuses from the gel and accumulates on the bottom of the wells. This high urea concentration affects separation of the RNA during gel run.

Acknowledgments

We thank Sabine Rottmüller and Bernd Haas for technical assistance, Elisabeth Kremmer for antibody production, and Vladimir Benes for affymetrix array analysis. This work was in part supported by a grant from the Deutsche Forschungsgemeinschaft (DFG, FO855), the European Union (LSHG-CT-2006-037900), and the Max-Planck-Society.

References

1. Bartel, D. P. (2009) MicroRNAs: target recognition and regulatory functions. *Cell* **136**, 215–33.
2. Carthew, R. W., and Sontheimer, E. J. (2009) Origins and Mechanisms of miRNAs and siRNAs. *Cell* **136**, 642–55.
3. Chen, P. Y., and Meister, G. (2005) microRNA-guided posttranscriptional gene regulation. *Biol Chem* **386**, 1205–18.
4. Meister, G., and Tuschl, T. (2004) Mechanisms of gene silencing by double-stranded RNA. *Nature* **431**, 343–9.
5. Parker, J. S., and Barford, D. (2006) Argonaute: A scaffold for the function of short regulatory RNAs. *Trends Biochem Sci* **31**, 622–30.
6. Peters, L., and Meister, G. (2007) Argonaute proteins: mediators of RNA silencing. *Mol Cell* **26**, 611–23.
7. Tolia, N. H., and Joshua-Tor, L. (2007) Slicer and the argonautes. *Nat Chem Biol* **3**, 36–43.
8. Jinek, M., and Doudna, J. A. (2009) A three-dimensional view of the molecular machinery of RNA interference. *Nature* **457**, 405–12.
9. Chen, K., and Rajewsky, N. (2007) The evolution of gene regulation by transcription factors and microRNAs. *Nat Rev Genet* **8**, 93–103.
10. Beitzinger, M., Peters, L., Zhu, J. Y., Kremmer, E., and Meister, G. (2007) Identification of human microRNA targets from isolated argonaute protein complexes. *RNA Biol* **4**, 76–84.
11. Easow, G., Teleman, A. A., and Cohen, S. M. (2007) Isolation of microRNA targets by miRNP immunopurification. *RNA* **13**, 1198–204.
12. Hendrickson, D. G., Hogan, D. J., Herschlag, D., Ferrell, J. E., and Brown, P. O. (2008) Systematic identification of mRNAs recruited to argonaute 2 by specific microRNAs and corresponding changes in transcript abundance. *PLoS ONE* **3**, e2126.
13. Karginov, F. V., Conaco, C., Xuan, Z., Schmidt, B. H., Parker, J. S., Mandel, G., and Hannon, G. J. (2007) A biochemical approach to identifying microRNA targets. *Proc Natl Acad Sci U S A* **104**, 19291–6.
14. Landthaler, M., Gaidatzis, D., Rothballer, A., Chen, P. Y., Soll, S. J., Dinic, L., Ojo, T., Hafner, M., Zavolan, M., and Tuschl, T. (2008) Molecular characterization of human Argonaute-containing ribonucleoprotein complexes and their bound target mRNAs. *RNA* **14**, 2580–96.
15. Weinmann, L., Hock, J., Ivacevic, T., Ohrt, T., Mutze, J., Schwille, P., Kremmer, E., Benes, V., Urlaub, H., and Meister, G. (2009) Importin 8 is a gene silencing factor that targets argonaute proteins to distinct mRNAs. *Cell* **136**, 496–507.
16. Rudel, S., Flatley, A., Weinmann, L., Kremmer, E., and Meister, G. (2008) A multifunctional human Argonaute2-specific monoclonal antibody. *RNA* **14**, 1244–53.
17. Pall, G. S., and Hamilton, A. J. (2008) Improved northern blot method for enhanced detection of small RNA. *Nat Protoc* **3**, 1077–84.

Chapter 13

Comprehensive Identification of miRNA Target Sites in Live Animals

Dimitrios G. Zisoulis, Gene W. Yeo, and Amy E. Pasquinelli

Abstract

MicroRNAs (miRNAs) are small RNA molecules that posttranscriptionally regulate the expression of protein-coding genes. The mature miRNAs are loaded into Argonaute-containing protein complexes (miRISC, *miR*NA *I*nduced *S*ilencing *C*omplex), and guide these complexes to the 3′ UTR of targeted mRNA transcripts via base-pairing interactions. However, the imperfect complementarity that characterizes the interactions between animal miRNAs and target sites complicates the identification of direct target genes. We developed a biochemical method to identify on a large scale the target sequences recognized by miRISC in vivo. The mRNA sites bound by miRISC are stabilized by cross-linking and isolated by immunoprecipitation of Argonaute-containing complexes. The bound RNA molecules are trimmed to the regions protected by Argonaute, subjected to a series of isolation and linker ligation steps and identified by high-throughput sequencing methods.

Key words: MicroRNAs, Argonaute Protein, CLIP-seq, Parallel sequencing, miRNA target sequences, Posttranscriptional gene regulation

1. Introduction

MicroRNAs (miRNAs) represent a newly discovered family of small, genomically encoded RNAs that posttranscriptionally regulate the expression of protein-coding genes (1). miRNAs guide Argonaute-containing protein complexes by base-pairing to the 3′ untranslated regions (3′UTRs) of target mRNAs, leading to their degradation and/or translational repression. In plants, miRNAs typically base-pair with near perfect complementary to their target 3′UTRs, but in contrast, most animal miRNAs base-pair only partially to their targets. Despite the intense efforts by the miRNA research community, elucidation of the miRNA target recognition mechanism remains challenging: the imperfect

Tamas Dalmay (ed.), *MicroRNAs in Development: Methods and Protocols*, Methods in Molecular Biology, vol. 732,
DOI 10.1007/978-1-61779-083-6_13,

complementarity between miRNAs and their target mRNAs, along with the limited available data describing these interactions, complicates the identification of target genes. Furthermore, computational analyses of miRNA targets that rely on the limited set of miRNA/mRNA target interactions that have been validated largely by heterologous reporter gene assays may result in physiologically irrelevant predictions.

Through the systematic isolation of miRISC-associated mRNA transcripts, a number of elegant studies have identified endogenous transcripts subject to miRNA-mediated regulation (2–6). However, to gain insight into the mechanism of miRNA target recognition, the specific sites of interaction between miRNAs and target mRNAs need greater resolution. To meet this challenge, we developed a comprehensive strategy to capture and identify miRNA target sequences by adapting the CLIP-seq (Cross-linking immunoprecipitation with high-throughput sequencing) method in *Caenorhabditis elegans* (7–9). A recent application of this approach in mouse brain resulted in a map of sites bound by Argonaute (Ago) in this tissue (10). *C. elegans* provided several distinct advantages for applying the CLIP-seq method for the identification of endogenous miRISC-binding sites: (a) a single Argonaute protein, *A*rgonaute-*L*ike *G*ene 1 (ALG-1), is largely responsible for miRNA function; (b) a viable *alg-1* genetic mutant control allowed us to eliminate background sequence noise extensively; (c) a short but well-established list of miRNA targets expected to be bound by ALG-1 at discrete positions is available and can be used to assess the sensitivity of the method; and (d) the in vivo binding context of endogenous ALG-1:RNA interactions in live animals is preserved.

We developed a procedure to identify Argonaute-binding sites in *C. elegans* by adapting the CLIP-seq method from the Darnell and Yeo labs (7, 8). Briefly, the ALG-1 CLIP-seq strategy involves in vivo crosslinking of miRNAs and mRNAs to the Argonaute protein ALG-1, thus stabilizing the interactions between miRNAs and their targets with the ALG-1 protein. The ALG-1-bound RNA molecules are isolated by immunoprecipitating the Argonaute protein and then subjected to trimming with an endonuclease. The refined ALG-1-bound RNA sequences are next purified and ligated to RNA adapters facilitating amplification and subsequent analysis by high-throughput sequencing. This approach should be adaptable for the identification of RNA sequences bound by other proteins in *C. elegans*, although we recommend that a robust antibody or rescuing tagged transgenic protein as well as mutant controls be utilized. Here, we describe the ALG-1 CLIP-seq method as applied to the identification of functional sites of miRNA::mRNA interactions in vivo in the *C. elegans* system (11).

2. Materials

2.1. Worm Cultures, UV Cross-linking, and Homogenization

Reagents and equipment used in multiple steps of the procedure are only mentioned the first time they are encountered.

1. Suitable materials for growing worms (OP50-seeded LB agar plates, etc.).

   M9 buffer: 22 mM KH_2PO_4, 22 mM Na_2HPO_4, 85 mM NaCl, and 1 mM $MgSO_4$. Autoclave and store at room temperature.
2. UV Crosslinker (Spectrolinker XL-1000, Spectronics Corporation).
3. Standard LB-Agar plates for worms with no bacterial food on them.
4. Homogenization buffer: 100 mM NaCl, 25 mM HEPES, 250 μM EDTA, 2 mM dithiothreitol (DTT), 0.1% (w/v) NP-40 (Amersham Biosciences), 25 U/ml RNAsin (Promega), and protease inhibitors (Complete Mini, Roche). All stock salt solutions should be autoclaved or sterile-filtered. Stock solutions of DTT (1 M) should be stored at −20°C in small aliquots, while NP-40 (10% solution) should not be used if older than a month. The homogenization buffer should be made fresh, and RNAsin and protease inhibitors (1 tablet per 20 ml) should be added prior to using the buffer.
5. Sonic Dismembrator (Model 100, Fisher Scientific).
6. Refrigerated tabletop microcentrifuge (Eppendorf).

2.2. Immunoprecipitation

1. Protein-G Sepharose beads (GE Healthcare).
2. Shaker (3D gyratory Nutator).
3. Anti-ALG-1 Rabbit IgG Antibody (0.1 mg/ml, Custom Antibody, Open Biosystems). You may also need an isotype control antibody (i.e. an non-ALG-1 specific IgG antibody) if not using genetic mutants, e.g. *alg-1* (*gk214*) mutants.
4. Bradford Reagents for protein quantification.
5. Wash buffer: 1× PBS (20 mM Tris–HCl, pH 7.4, 137 mM NaCl), 0.1% (w/v) sodium dodecyl sulfate (SDS), 0.5% (w/v) sodium deoxycholate, and 0.5% (w/v) NP-40. The PBS solution should not contain Mg^{+2} or Ca^{+2} as this may lead to RNA degradation. Store at room temperature without NP-40, which should be added prior to use.
6. High salt wash buffer: 5× PBS, 0.1% (w/v) SDS, 0.5% (w/v) sodium deoxycholate, and 0.5% (w/v) NP-40. As before, Mg^{+2} or Ca^{+2} should not be present in the PBS solution. Sterile-filter and store at room temperature without NP-40, which should be added prior to use.

7. Polynucleotide kinase buffer (PNK buffer): 50 mM Tris–HCl, pH 7.4, 10 mM $MgCl_2$ and 0.5% (w/v) NP-40. Sterile-filter and store at room temperature without NP-40, which should be added prior to use.

2.3. RNA Trimming

1. Micrococcal nuclease (2×10^3 U/μl, NEB) and micrococcal nuclease (MN) buffer (50 mM Tris–HCl, pH 7.9, 5 mM $CaCl_2$). Aliquot and store at −20°C.
2. Adjustable temperature shaker (Thermomixer R, Eppendorf).
3. PNK+EGTA buffer: 50 mM Tris–HCl, pH 7.4, 20 mM EGTA and 0.5% (w/v) NP-40. The EGTA stock solution (0.5 M) should be adjusted to pH 8 for full solubility. Buffer may be stored at room temperature without NP-40, which should be added prior to use.

2.4. Alkaline Phosphatase Treatment

1. Calf-intestine alkaline phosphatase (CIP, 10 U/μl, NEB) and 1× enzyme buffer #3 (NEB).
2. 0.01% (w/v) (or 0.1 mg/ml) Bovine serum albumin solution (BSA, NEB) in DEPC-treated dH_2O.

2.5. 3′ RNA Adapter Ligation

1. 20 μM 3′ RNA adapter (5′-UCG UAU GCC GUC UUC UGC UUG-3′) with a puromycin modification at the 3′ end (PAGE-purified, Dharmacon).
2. T4 RNA Ligase Kit (Fermentas) containing T4 RNA ligase, T4 RNA ligase buffer, BSA 1 mg/ml and 10 mM ATP.

2.6. Polynucleotide Kinase Treatment

1. T4 Polynucleotide kinase enzyme (PNK, NEB) and the 10× PNK buffer (NEB).
2. ^{32}P-γ-ATP (6,000 Ci/mmol, 150 mCi/ml, 5 mCi total, Perkin Elmer).

2.7. SDS-PAGE, Transfer to Nitrocellulose Membrane and Western Blot Analysis

1. Sample buffer: Nupage LDS 4× (Invitrogen), without any reducing agents.
2. Full range protein ladder (Rainbow Ladder, GE Healthcare).
3. 10% Bis-Tris gel (Invitrogen).
4. Running buffer: MOPS SDS running buffer (Invitrogen).
5. Nitrocellulose membrane (0.45-μm pore size, Biorad).
6. Transfer buffer: Nupage transfer buffer (Invitrogen), supplemented with 5% (v/v) methanol and 0.01% (w/v) SDS.
7. Wet transfer apparatus (Biorad).
8. MR autoradiogram film (Kodak).
9. Tris-buffered saline (TBS)-Tween (T): 20 mM Tris–HCl, pH 7.6, 137 mM NaCl, and 0.1% (v/v) Tween 20.
10. Blocking buffer: 5% (w/v) nonfat dry milk in TBS-T buffer.
11. Secondary antibody: horseradish-peroxidase-conjugated anti-rabbit IgG (0.8 mg/ml, Thermo Scientific).

12. Enhanced chemiluminescence reagents (ECL Plus, GE Healthcare).

2.8. RNA Isolation and Purification

1. Proteinase K (Invitrogen) and Proteinase K (PK) buffer: 100 mM Tris–HCl, pH 7.5, 50 mM NaCl, and 10 mM EDTA. Store at room temperature.
2. 7 M Urea solution in PK buffer. This solution must be made fresh prior to use by weighing 0.42 g of urea and adding PK buffer to 1 ml.
3. RNA phenol (Fisher Scientific). Store at 4°C where it freezes. It should be thawed and brought to room temperature prior to use.
4. Chloroform:isoamyl solution (49:1, Fisher Scientific).
5. 3 M Sodium acetate, pH 5.2.
6. Glycogen, 20 mg/ml (Ambion). Make single-use aliquots and store at –20°C.
7. Ethanol:isopropanol solution 1:1 (v/v). Store at room temperature.
8. 75% (v/v) ethanol in DEPC-treated dH_2O. Store at 4°C.

2.9. 5′ RNA Adapter Ligation and DNAse Treatment

1. 20 μM 5′ RNA adapter (5′-GUU CAG AGU UCU ACA GUC CGA CGA UC-3′) with a biotin modification at the 5′ end (PAGE-purified, Dharmacon).
2. T4 RNA Ligase kit (Fermentas) containing T4 RNA ligase, T4 RNA ligase buffer, BSA 1 mg/ml, and 10 mM ATP.
3. 10× DNAse I buffer: 100 mM Tris–HCl, pH 7.6, 25 mM $MgCl_2$, and 5 mM $CaCl_2$.
4. RQ1 DNAse (Promega).

2.10. Reverse Transcription and Polymerase Chain Reaction

1. Primers: 20 μM 3′ PCR Primer (5′-CAA GCA GAA GAC GGC ATA CGA-3′) and 20 μM 5′ PCR primer (5′-AAT GAT ACG GCG ACC ACC GAC AGG TTC AGA GTT CTA CAG TCC GA-3′). Primers are PAGE-purified (IDT DNA).
2. Superscript III Reverse Transcription kit (Invitrogen) containing 10 mM dNTPs, 0.1 M DTT, reverse transcription buffer (first-strand synthesis buffer), and SuperScript III enzyme.
3. Phusion PCR kit (NEB) containing 5× DNA Phusion polymerase buffer, 10 μM dNTPs, and Phusion polymerase.
4. Thermocycler (Eppendorf).

2.11. Agarose Gel Analysis of PCR Product and Gel Extraction

1. 25× Tris-Acetate-EDTA buffer (TAE, Fisher Scientific).
2. Agarose (low melting point, molecular biology grade, Fisher Scientific).

3. 1× Tris-EDTA buffer (TE): 10 mM Tris–HCl, pH 7.5, 1 mM EDTA.
4. QIAEX II Gel extraction system (QIAGEN).
5. Standard lab equipment for casting and running Agarose gels.
6. 10 mg/ml ethidium bromide.
7. 6x DNA loading buffer (Invitrogen).
8. 25–bp DNA ladder (Invitrogen).
9. Heavy duty sheet protectors (Office Supplies).

3. Methods

The key steps are outlined in Fig. 1.

3.1. UV Cross-linking of C. elegans and Lysis

Live *C. elegans* animals are subjected to UV irradiation in order to cross-link proteins and RNA molecules, allowing the ALG-1 to be cross-linked to the bound mRNAs and miRNAs. The collected worms are then lysed by sonication, effectively disrupting the worm cuticle and cell membranes.

1. Grow worms as usual and collect the desired stage(s), ~50,000 worms or enough (~50 µl of worms) to obtain ~1 mg of total protein per sample.
2. Wash worms (3×) with M9 in a 15–ml falcon tube and rock for 20 min to allow the worms to digest the bacterial food present in their intestine.
3. Collect the worms and resuspend in 1 ml of M9 buffer.
4. Plate the worms on worm plates (100 mm) with no food at room temperature (RT) as you would plate bacteria on a plate. Make sure the worms are in one layer (avoid overlapping worms or areas of high concentration where worms can burrow under each other).
5. Place the worm plates into the Spectrolinker XL-1000 (without the plate lids) and irradiate with UV-B at 3 kJ/m^2 (Power Settings: 3,000). Collect the worms in M9 buffer (they should all be immobilized now) and resuspend in 1 ml of M9 and transfer to a 2-ml Eppendorf tube (round bottom).
6. Spin down for 30 s at 3,000 × *g* and resuspend the worms in 700 µl of ice-cold homogenization buffer. From now on, keep the worms on ice.
7. Sonicate the worms with 10-s pulses (5×) and resting for 50 s on ice between sonications (~18 W RMS output power on the Dismembrator).

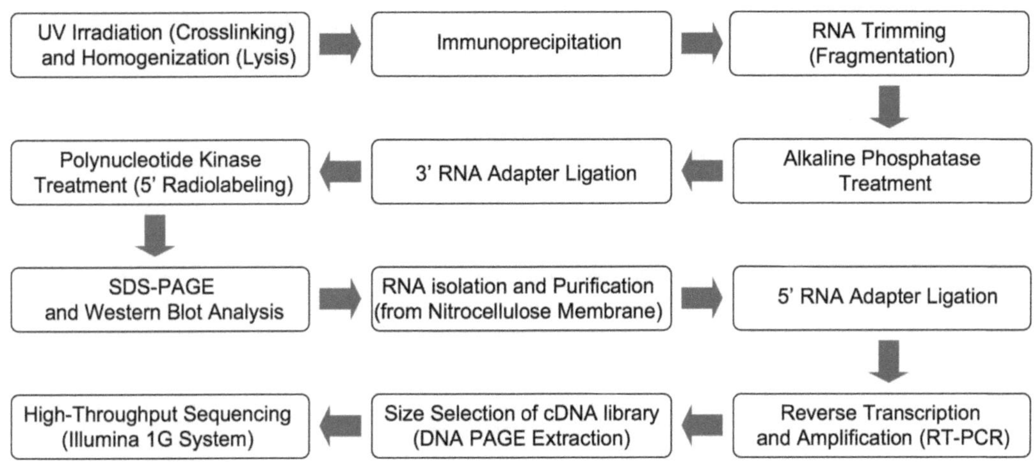

Fig. 1. Outline of the ALG-1 CLIP-seq experimental approach: i) *in vivo* stabilization of the sites of interaction between miRNAs, target mRNAs and the Argonaute protein ALG-1 by UV crosslinking; ii) immunoprecipitation with an ALG-1 specific antibody under stringent wash conditions; iii) trimming of the RNA molecules (mRNA and miRNAs) that are crosslinked to the ALG-1-protein by the endonuclease micrococcal nuclease (MNase) resulting in short fragments, the majority of which are protected from further degradation by the ALG-1-containing miRISC complex; iv) removal of the residual 3′ phosphoryl group, which was created by the MNase digestion, by phosphatase treatment allowing, thus, the ligation reaction in the next step; v) ligation of an RNA adapter to the 3′ hydroxyl group of the RNA fragments (the adapter has a 3′-puromycin modification to avoid self-circularization and subsequent concatemer formation); vi) labeling of the 5′ end by polynucleotide kinase with ^{32}P-γ-ATP, which also converts the residual 5′-hydroxyl group to 5′phosphoryl, allowing for the ligation of an adapter at a later stage; vii) native SDS-PAGE in order to isolate the band corresponding to the crosslinked ALG-1/RNA complexes and Western blot analysis to verify that the appropriate bands were isolated; viii) proteinase degradation of the ALG-1 protein and total RNA extraction; ix) ligation with a 5′ biotin-modified RNA adapter to avoid self-circularization and concatemer formation; x) Reverse Transcription and PCR amplification; xi) agarose gel analysis and size selection of the libraries and finally; xii) high-throughput sequencing using the Illumina 1G system.

8. Centrifuge lysates at 16,000×*g* for 15 min at 4°C (refrigerated microcentrifuge) and collect supernatant. Lysates can be stored at −80°C.

3.2. Immuno-precipitation

The complexes containing Argonaute and the crosslinked mRNAs/miRNAs are immunoprecipitated, with the use of the polyclonal Rabbit Antibody specific for ALG-1, under stringent conditions (11) (see Note 1).

1. Wash the Protein G–sepharose beads for 30 s at 1,000×*g* with homogenization buffer (3×). You will need 170 μl of beads (50:50 slurry) for every sample. All washes and centrifugations involving beads are performed for 30 s at 1,000×*g*.
2. Resuspend the beads (~85 μl) in equal volume of homogenization buffer (~85 μl) for 170 μl of 50:50 slurry.

3. Quantify the protein content of the samples by Bradford Assay and adjust protein concentration of the lysate supernatants to 1 mg/ml. Save 20 μl for Western Blot Analysis (Input).
4. Preclear lysates by adding 50 μl of the washed beads slurry (50:50) for 1 h at 4°C with gentle shaking on a gyratory Nutator.
5. Centrifuge tubes and collect supernatant. Add antibody and incubate overnight with gentle shaking at 4°C (7 μg/ml for Rabbit anti-ALG-1).
6. Add 100 μl of washed bead slurry (50:50) and incubate at 1 h at 4°C with gentle shaking to allow the beads to bind to the antibody.
7. Collect the beads by centrifugation and save 20 μl of the supernatant for Western Blot Analysis (Supernatant).
8. Wash the beads with ice-cold wash buffer (2×).
9. Wash the beads with ice-cold high-salt wash buffer (2×).
10. Wash the beads with ice-cold PNK buffer (2×).

3.3. RNA Trimming

The ALG-1-bound RNAs are trimmed by micrococcal nuclease, resulting in random-length sequences around a central RNA region protected by Argonaute. This endonuclease leaves 3′ phosphate and 5′ hydroxyl groups, which are essential for further biochemical modifications and can be inactivated by EGTA, preventing RNA degradation that could compromise RNA recovery in subsequent steps.

1. Dilute Micrococcal Nuclease to 1:100 (20 U/μl) in MN buffer (see Note 2).
2. Add 499 μl of MN buffer to the washed beads and 1 μl of the 1:100 micrococcal nuclease solution.
3. Incubate at 4°C for 10 min with intermittent shaking in a Thermomixer R (1,200 rpm for 1 min and then 1,200 rpm for 15 s every 3 min).
4. Wash the beads with ice-cold PNK+EGTA buffer (2×). EGTA is used to inactivate the micrococcal nuclease.
5. Wash the beads with ice-cold wash buffer (2×).
6. Wash the beads with ice-cold PNK buffer (2×).

3.4. Alkaline Phosphatase Treatment

The 3′ phosphate group of the fragmented RNA tag is removed by alkaline phosphatase, enabling the ligation of the 3′ RNA adapter at the next step.

1. Prepare CIP mix:

8 μl	10× NEB buffer #3
3 μl	CIP
69 μl	dH_2O
80 μl	Total

2. Add CIP mix to the washed beads and incubate at 37°C for 10 min with intermittent shaking (1,200 rpm for 15 s every 3 min).
3. Wash beads with PNK+EGTA buffer (2×).
4. Wash beads with PNK buffer (2×).
5. Wash beads with BSA solution (0.1 mg/ml of BSA in DEPC treated dH_2O). BSA enhances T4 RNA ligase efficiency and prevents bead aggregation.

3.5. 3′ RNA Adapter Ligation

The 3′ RNA adapter is ligated to the ALG-1-bound mRNAs/miRNAs by the T4 RNA ligase. The adapter is PAGE-purified and has a puromycin modification at the 3′ end to avoid self-circularization. The adapters allow the PCR amplification and high-throughput sequencing of RNA fragments in later steps.

1. Prepare adapter/ligase mix:

8 μl	3′ RNA adapter (20 μM)
8 μl	10× T4 RNA ligase buffer
1.6 μl	BSA (1 mg/ml)
8 μl	ATP (10 mM)
4 μl	T4 RNA ligase
49.4 μl	dH_2O
1 μl	RNAsin
80 μl	Total

2. Add adapter/ligase mix to the washed beads.
3. Incubate at 16°C overnight with intermittent shaking (1,300 rpm for 15 s every 5 min).
4. Prepare new adapter/ligase mix:

5 μl	3′ RNA adapter (20 μM)
2 μl	T4 RNA ligase
4 μl	ATP (10 mM)
11 μl	Total

5. Add to beads with the previous adapter/ligase mix and incubate for another 3 h as before.
6. Wash with PNK buffer (3×).

3.6. Polynucleotide Kinase Treatment

At this step, the 5′ hydroxyl of the RNA tag due to the micrococcal nuclease treatment is converted into a 5′ phosphate group by the PNK. This allows the ligation of the 5′ RNA adapter later. Furthermore, use of ^{32}P-γ-ATP enables the effective radio-labeling of the ALG-1/RNA complexes for identification during the SDS-PAGE step.

1. Prepare PNK mix:

8 μl	10× PNK buffer
2 μl	P^{32}-γ-ATP (300 μCi total)
4 μl	T4 PNK enzyme
66 μl	dH_2O
80 μl	Total

2. Add the PNK mix to the washed beads.
3. Incubate at 37°C for 10 min with intermittent shaking (1,000 rpm for 15 s every 4 min).
4. Add 10 μl of 1 mM ATP and incubate for 5 min as before.
5. Wash with PNK+EGTA buffer (3×).

3.7. SDS-PAGE and Transfer to Nitrocellulose Membrane and Western Blot Analysis

The ALG-1/mRNA/miRNA complexes are eluted from the beads, loaded on a native gel and transferred onto a nitrocellulose membrane. Use of native PAGE conditions prevents protein denaturation and RNA degradation. Once transferred to the nitrocellulose membrane, the membrane piece corresponding to the diffuse radioactive band, which contains the ALG-1/RNA complexes, is isolated and the RNA is extracted. This step increases the sensitivity of the approach and reduces background contamination. Next, the membrane is subjected to Western Blot Analysis in order to verify that the appropriate band was isolated and that the immunoprecipitation was successful.

1. Resuspend the beads in 30 μl of PNK+EGTA buffer and 10 μl of 4× Nupage LDS sample buffer without any reducing agents.
2. Mix 20 μl of Input and Supernatant controls (step 3.2.3), as well as protein ladder with 7 μl of 4× LDS sample buffer.
3. Incubate at 70°C for 10 min at 1,000 rpm (controls, ladder and samples) and take the bead eluate (IP) for loading.
4. Load samples, controls, and ladder onto a native 10% Bis-Tris gel and run at 150 V, with MOPS running buffer, in a standard PAGE apparatus.

5. Transfer gel to a nitrocellulose membrane of 0.45-μm pore size using a standard wet transfer apparatus with transfer buffer at 40 V for 4 h. Make sure that the apparatus is cooled, and two Whatman papers are on both sides of the gel/membrane.
6. After transfer, rinse the nitrocellulose membrane with 1× PBS, gently blot on absorbing paper, wrap the membrane in plastic wrap, and expose to autoradiogram film (Fig. 2a). Under native conditions, the 150-kDa protein molecular marker separates into two bands and the middle of the band is marked as 150 kDa (Fig. 2). The ALG-1/RNA complexes appear as a diffused radioactive band running at approximately the middle of protein marker bands, about 30 kDa higher than the expected molecular weight of ALG-1 in a denaturing gel.
7. Cut out the band corresponding to the ALG-1/RNA complexes with a clean razor and place into separate 1.5-ml Eppendorf tubes. For controls, including *alg-1* mutants or immunoprecipitation with control antibody (IgG), isolate the band that would correspond to that area, using the wild-type sample as a guide. It may help to place the developed autoradiogram film under the nitrocellulose membrane on a lightbox to mark the position of the band on the membrane. You may also count radioactivity in a scintillation counter. The nitrocellulose pieces can be stored at −80°C.
8. At this point, a standard Western blot analysis may be performed as a control for the immunoprecipitation efficiency and to verify that the correct bands were isolated in the previous step. Wash the nitrocellulose membrane with PBS.
9. Block the membrane with 10 ml of 5% nonfat milk in TBS-T buffer for 30 min at RT.
10. Incubate the membrane overnight at 4°C with the primary antibody (1:500 dilution of Rabbit anti-ALG-1 in 10 ml of 5% nonfat milk in TBS-T buffer) (see Note 3).
11. Wash membrane extensively with TBS-T for 10 min (3×).
12. Incubate the membrane at RT for 1 h with the secondary antibody (1:10,000 dilution of horseradish peroxidase-conjugated anti-rabbit IgG in 10 ml of 5% nonfat milk in TBS-T buffer) (see Note 3).
13. Wash membrane extensively with TBS-T for 10 min (3×).
14. Visualize the bands on the nitrocellulose membrane using enhanced chemiluminescence reagents (ECL Plus, GE Healthcare). Remove excess TBS-T buffer by blotting gently on absorbing paper. Mix 3 ml of solution A and 75 μl of solution B at RT, cover the membrane, and incubate for 5 min at RT.
15. Remove excess ECL reagent by blotting gently on absorbing paper, wrap in plastic wrap, and expose to film (Fig. 2b).

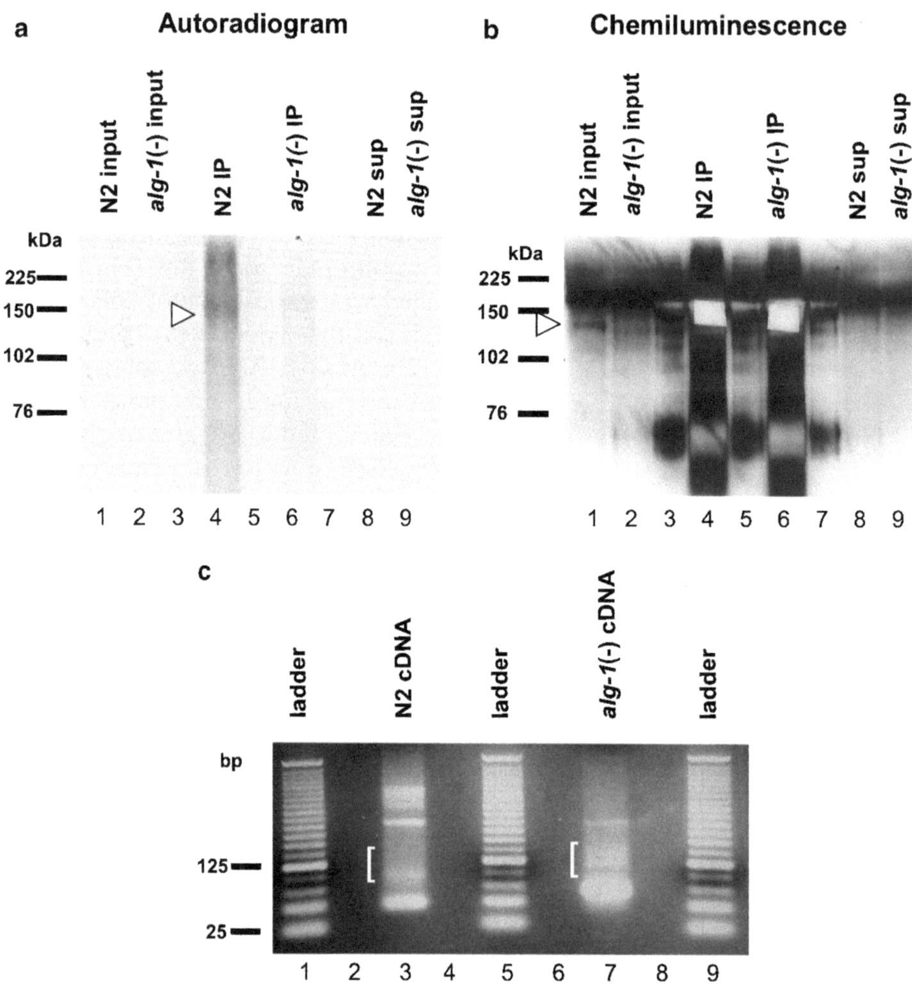

Fig. 2. Key stages of the ALG-1 CLIP-seq methodology. (**a**) Autoradiogram of the nitrocellulose membrane after SDS-PAGE. The *white arrowhead* indicates the diffuse radioactive band corresponding to the radio-labeled ALG-1/RNA complexes in the wild-type IP sample (N2). As a control, the C. elegans strain *gk214*, mutant for the *alg-1* gene, was used. Lanes 3, 5, and 7 are blank. (**b**). Western Blot analysis of the same membrane after the isolation of the nitrocellulose pieces containing the ALG-1/RNA complexes (**a**, N2) or background control (**a**, *alg-1*(−)). The *white arrowhead* indicates the ALG-1 band from the input control (N2). Note: (1) under native conditions, the Rainbow Protein Marker for 150 kDa separates into two and annotated 150-kDa marker corresponds to the middle of the distance of the two separate bands, (2) the radio-labeled ALG-1/RNA complex runs slower than input due to the adapter ligation that affects the associated RNA size, (3) the strong background present in the immunoprecipitation lanes (IP) is due to the fact that the same antibody was used for the immunoprecipitation and for Western blot analysis, (4) the absence of an ALG-1 band in the *alg-1*(−) mutants demonstrates the specificity of the antibody, and (5) the almost complete depletion of the ALG-1 protein from the supernatant controls. Lanes 3, 5, and 7 are blank. (**c**) Agarose gel analysis of the cDNA libraries prior to size selection. The white brackets indicate the range of cDNA sizes that was isolated and subjected to high-throughput sequencing (75–150 bp). Lanes 1, 5, and 9 contain the 25-bp DNA ladder, and lanes 2, 4, 6 and 8 are blank.

3.8. RNA Isolation and Purification

Next, the isolated nitrocellulose pieces containing the ALG-1/RNA complexes are subjected to protein degradation and the RNA is extracted and precipitated. This step degrades the ALG-1 protein and allows the previously ALG-1-bound RNA molecules to be isolated.

1. Make a 4 mg/ml proteinase K solution in PK buffer and incubate this solution at 37°C for 20 min to kill any RNAses that may be present.
2. Add 200 μl of proteinase K solution to each tube of isolated nitrocellulose pieces; incubate at 37°C for 20 min at 1,200 rpm.
3. Add 200 μl of freshly made 7 M urea/PK buffer solution and incubate at 37°C for 20 min at 1,000 rpm.
4. Add 400 μl RNA phenol and 130 μl of $CHCl_3$ (chloroform:isoamyl alcohol, 49:1) to the solution, mix vigorously by hand for a few seconds, and incubate at 37°C for 20 min at 1,000 rpm.
5. Mix the tubes vigorously by hand for a few seconds and centrifuge the tubes at 16,000×*g* for 5 min at RT.
6. Transfer the aqueous phase into a new Eppendorf tube and add:

50 μl	3 M Sodium acetate, pH 5.2
1 μl	Glycogen
1 ml	1:1 Ethanol:isopropanol solution

7. Precipitate overnight at –20°C or for longer than 1 h.

3.9. 5′ RNA Adapter Ligation and DNAse Treatment

The extracted RNA is resuspended and the 5′ RNA adapter is ligated to the RNA molecules with the T4 RNA Ligase. The adapter has a biotin modification at the 5′ end to avoid self-circularization. This allows the amplification of the RNA fragments in later steps of the method.

1. Centrifuge samples at 4°C, 16,000×*g* for 15 min. You may check radioactivity counts to ensure efficient precipitation.
2. Wash pellet with 300 μl of ice-cold 75% ethanol (2×). The washing steps remove residual salt, which may decrease ligation efficiency.
3. Centrifuge samples at 4°C, 16,000×*g* for 5 min.
4. Wash again with 150 μl of 75% ethanol, and this time you may displace the pellet by flicking the tube briefly.
5. Centrifuge samples at 4°C, 16,000×*g* for 10 min.

6. Air-dry samples or briefly dry in a speedvac (~2 min) making sure they do not overdry.
7. Check precipitation efficiency by radioactive counts.
8. Prepare ligation mix:

1 μl	10× T4 RNA ligase buffer
0.2 μl	BSA (1 mg/ml)
1 μl	ATP (10 mM)
1 μl	5′ RNA adapter (20 μM)
1 μl	RNAsin
5.3 μl	dH_2O
0.5 μl	T4 RNA ligase
10 μl	Total

9. Add to RNA pellets and resuspend gently by pipetting.
10. Incubate at 16°C for 1 h (or overnight). After treatment, do not heat inactivate the ligase: RNA will be degraded with Mg^{+2} present, while an EDTA inactivation would sequester Mg^{+2} necessary for the subsequent DNAse treatment.
11. Prepare DNAse mix:

79 μl	dH_2O
11 μl	10× DNAse I buffer
5 μl	RNAsin
5 μl	RQ1 DNAse
100 μl	Total

12. Add to RNA sample and incubate at 37°C for 20 min.
13. Add:

300 μl	dH_2O
300 μl	RNA phenol
100 μl	$CHCl_3$ (chloroform:isoamyl alcohol, 49:1)

14. Vortex briefly and centrifuge at 16,000×*g* for 5 min.
15. Remove supernatant into a new tube and add:

50 μl	3 M Sodium acetate, pH 5.2
1 μl	Glycogen
1 ml	1:1 Ethanol:isopropanol solution

16. Precipitate at –20°C overnight or longer than 1 h.

3.10. Reverse Transcription and Polymerase Chain Reaction

The precipitated RNA fragments are reverse transcribed by using a primer complementary to the ligated 3′ RNA linker. Using the suitable primers for both 5′ and 3′ ends, the cDNA molecules are amplified by polymerase chain reaction.

1. Centrifuge samples at 4°C, 16,000×*g* for 15 min. Check radioactivity counts to ensure efficient precipitation.
2. Wash pellet with 300 μl of ice-cold 75% ethanol (2×).
3. Air-dry samples or briefly dry in a speedvac (~2 min) making sure they do not overdry.
4. Resuspend in 9.5 μl of dH_2O.
5. Add 0.5 μl of 20 μM 3′ PCR Primer and mix by pipetting.
6. Heat at 65°C for 5 min, chill for 1 min on ice and quick spin.
7. Make a reverse transcription mix:

1 μl	10 mM dNTPs
1 μl	0.1 M DTT
4 μl	5× SuperScript RT buffer
1 μl	RNAsin
1 μl	SuperScript III
2 μl	dH_2O
10 μl	Total

8. Add reverse transcription mix, and mix gently by pipetting.
9. Incubate at 50°C for 30 min, 90°C for 5 min and leave at 4°C. Store 10 μl of the RT product at −20°C and use 10 μl for the PCR step.
10. Perform PCR using the NEB Phusion Kit. Prepare PCR mix by adding:

6 μl	5× DNA Phusion polymerase buffer
0.75 μl	20 μM 5′ PCR primer
0.75 μl	20 μM 3′ PCR primer
0.6 μl	10 μM dNTPs
0.5 μl	Phusion polymerase
11.4 μl	dH_2O
20 μl	Total

11. Add 20 μl of the PCR mix to 10 μl of the RT product and perform a PCR reaction with the following settings:

    98°C for 30 s

    35 cycles: 98°C for 10 s, 60°C for 30 s, and 72°C for 15 s

    72°C for 10 min

3.11. Agarose Gel Analysis of PCR Product and Gel Extraction

The PCR-amplified cDNA libraries are selected for sizes 75–150 bp on a standard Agarose gel. The size corresponds to the length of the isolated cDNA plus the length of the PCR primers and should be adjusted if linkers or primers of different sizes are used. This isolation step allows control over the sizes of molecules that will actually be sequenced and removes primer-dimer PCR artifacts. For the gel extraction, we use the Qiaex II silica particles, which bind DNA (40 bp and above) in solubilized agarose gel. Avoid using silica-membrane-based gel extraction systems that purify DNA sizes of 100 bp and above, since this will lead to loss of PCR products of smaller size.

1. Prepare a 3% Agarose gel in 1× TAE buffer, containing 0.5 μg/ml ethidium bromide, using standard lab equipment. Allow gel to polymerize for approximately 40 min and load samples in standard DNA loading buffer. Include 2 μl of 25 bp DNA ladder.
2. Run at 120–150 V in 1 × TAE buffer.
3. Place gel on a sheet protector and on the UV-transilluminator box cut the bands corresponding to sizes 75–150 nt (Fig. 2c).
4. Extract DNA from the gel using Qiaex II silica particles per manufacturer's instructions. For the final elution step, elute the DNA from the silica particles twice, using 100 μl of 1 × TE buffer and combine the two eluates for a total of 200 μl.
5. Precipitate by adding 1 μl of 20 mg/ml glycogen, 20 μl of 3 M sodium acetate (pH 5.2) and 500 μl of ethanol. Precipitate in dry ice-ethanol bath for 30 min or at –20°C overnight.
6. Centrifuge at 4°C for 20 min at 16,000 g. Wash pellet with 80% ethanol, dry pellet by speedvac and resuspend in 20 μl of dH_2O.

The DNA libraries are now ready to be analyzed with the Illumina 1G Sequencing system. For extra amplification steps and quality controls see Note 4.

4. Notes

1. It may be necessary to confirm that the antibody can indeed be used for immunoprecipitations. To achieve this, after completion of the immunoprecipitation and the washing steps, proceed to Subheading 3.7.
2. Optimization of the MNase amount maybe necessary for some applications. You may use different dilutions, ranging from 1:10 to 1:1,000 (2–200 U/μl) and proceed normally.
3. The dilutions of the primary and secondary antibody may need to be optimized to obtain clean results. In this method, we are using the only available anti-ALG-1 antibody for both immunoprecipitation and Western blot analysis, which results in the extra dark bands present in the Western blot. The use of antibodies raised in different systems should give better visual results.
4. If necessary, PCR may be performed again on the extracted CLIP-seq libraries as before, for 25 cycles or less. As a quality control, a sample of the CLIP-seq libraries may be cloned using the TOPO-cloning kit (Invitrogen) and sequenced to verify that adapter Concatemers is the right term.

References

1. Bartel, D. P. (2009) MicroRNAs: target recognition and regulatory functions. *Cell* **136**, 215–33.
2. Hendrickson, D. G., Hogan, D. J., Herschlag, D., Ferrell, J. E., and Brown, P. O. (2008) Systematic identification of mRNAs recruited to argonaute 2 by specific microRNAs and corresponding changes in transcript abundance. *PLoS ONE* **3**, e2126.
3. Brennecke, J., Stark, A., Russell, R. B., and Cohen, S. M. (2005) Principles of microRNA-target recognition. *PLoS Biol* **3**, e85.
4. Karginov, F. V., Conaco, C., Xuan, Z., Schmidt, B. H., Parker, J. S., Mandel, G., and Hannon, G. J. (2007) A biochemical approach to identifying microRNA targets. *Proc Natl Acad Sci USA* **104**, 19291–6.
5. Zhang, L., Ding, L., Cheung, T. H., Dong, M. Q., Chen, J., Sewell, A. K., Liu, X., Yates, J. R., 3rd, and Han, M. (2007) Systematic identification of C. elegans miRISC proteins, miRNAs, and mRNA targets by their interactions with GW182 proteins AIN-1 and AIN-2. *Mol Cell* **28**, 598–613.
6. Beitzinger, M., Peters, L., Zhu, J. Y., Kremmer, E., and Meister, G. (2007) Identification of human microRNA targets from isolated argonaute protein complexes. *RNA Biol* **4**, 76–84.
7. Licatalosi, D. D., Mele, A., Fak, J. J., Ule, J., Kayikci, M., Chi, S. W., Clark, T. A., Schweitzer, A. C., Blume, J. E., Wang, X., Darnell, J. C., and Darnell, R. B. (2008) HITS-CLIP yields genome-wide insights into brain alternative RNA processing. *Nature* **456**, 464–9.
8. Yeo, G. W., Coufal, N. G., Liang, T. Y., Peng, G. E., Fu, X. D., and Gage, F. H. (2009) An RNA code for the FOX2 splicing regulator revealed by mapping RNA-protein interactions in stem cells. *Nat Struct Mol Biol.* **16**, 130–7.
9. Sanford, J. R., Wang, X., Mort, M., Vanduyn, N., Cooper, D. N., Mooney, S. D., Edenberg, H. J., and Liu, Y. (2009) Splicing factor SFRS1 recognizes a functionally diverse landscape of RNA transcripts. *Genome Res* **19**, 381–94.
10. Chi, S. W., Zang, J. B., Mele, A., and Darnell, R. B. (2009) Argonaute HITS-CLIP decodes microRNA-mRNA interaction maps. *Nature* **460**, 479–86.
11. Zisoulis, D. G., Lovci, M. T., Wilbert, M. L., Liang, T. Y., Yeo, G. W., and A.E., Pasquinelli. (2010) Comprehensive discovery of endogenous Argonaute binding sites in *Caenorhabditis elegans Nat Struct Mol Biol.* **17,** 173–9.

Chapter 14

Target Validation of Plant microRNAs

César Llave, José Manuel Franco-Zorrilla, Roberto Solano, and Daniel Barajas

Abstract

microRNAs (miRNAs) regulate gene expression through sequence-specific interactions with cognate mRNAs that result in translational inhibition, mRNA decay, or slicing within the region of complementarity. miRNA processing activity on complementary target mRNAs generates 3′ end cleavage products that contain ligation-competent, 5′-monophosphates. Precise mapping of miRNA-directed cleavage sites within target transcripts is, therefore, possible using RNA ligase-mediated 5′ amplification of cDNA ends (RLM-RACE). Here, we provide a comprehensive RLM-RACE-based protocol for the amplification of 5′ ends derived from cleaved transcripts resulting from miRNA-guided cleavage events. Novel strategies for high-throughput analysis of miRNA cleavage products have emerged as powerful tools for the novo identification of miRNA targets in a genomic perspective. In this work, we also describe a novel methodology for genome-wide identification of miRNA targets that exploits RLM-RACE for non-sequence-specific enrichment of cleaved transcripts, T7 RNA polymerase-mediated amplification of target products, and microarray hybridization.

Key words: RLM-RACE, microRNAs, microRNA targets, Gene regulation, Microarray hybridization

1. Introduction

MicroRNAs (miRNAs) in plants account for about 15% of the total small RNA population and regulate the expression of a diverse set of genes at the posttranscriptional level (1, 2). Beyond their role in development, miRNAs control key aspects of hormone and immune responses and adaptation to a variety of biotic and abiotic stimuli in plants (2). miRNAs are produced by a double-stranded (dsRNA)-specific RNase III-like enzyme called Dicer-like1 (DCL1; there are four *DCL* genes in *Arabidopsis thaliana*) and recruited by Argonaute (AGO) proteins into the RNA-induced silencing effector complex (3). miRNAs guide repression of the target transcripts by any of several mechanisms that entail a combination of

Tamas Dalmay (ed.), *MicroRNAs in Development: Methods and Protocols*, Methods in Molecular Biology, vol. 732, DOI 10.1007/978-1-61779-083-6_14, © Springer Science+Business Media, LLC 2011

translational inhibition, mRNA decay, and slicing within the region of complementarity (4–6).

Plant miRNAs use single, highly matched target sites within their target mRNAs to promote slicing (4, 6). miRNA-guided, sequence-specific endonucleolytic cleavage events can be identified in vivo using RNA ligase-mediated 5′ rapid amplification of cDNA ends (RLM-RACE) (4, 7). miRNA processing activity on cognate target mRNAs generates 3′ end cleaved products that are relatively stable and contain ligation-competent, 5′-monophosphate ends rather than conventional 5′ caps (4). This feature makes it possible to ligate directly an RNA oligonucleotide adapter to the 5′ terminus of the 3′ end cleavage product using T4 RNA ligase, and without any further enzymatic pretreatment such as CIP- or TAP-treatments usually required for classical 5′ RLM-RACE methods (7, 8). The ligated RNA is then reverse transcribed into first-strand cDNA and PCR amplified using a reverse gene-specific primer (GSP) in combination with a forward primer homologous to the RNA adapter sequence (adapter-specific primer, ASP). The precise cleavage site, which is predicted to occur in the middle of the region of complementarity between the miRNA and the target mRNA, is given by the junction of the RNA adapter and the miRNA cleavage product.

High-throughput identification of mRNAs that are targeted for endonucleolytic cleavage by miRNAs and small interfering RNAs (siRNAs) can be achieved using a novel methodology that combines selective enrichment of 3′ end cleavage products using RLM-RACE, T7 RNA polymerase-mediated amplification of ligated products, and microarray hybridization (9). This novel methodology is not suitable for precise mapping of cleavage sites within target mRNAs, nevertheless it offers a valuable tool for "blind" identification of potential targets of miRNAs and siRNAs at a genomic scale (9).

2. Materials

2.1. Protocol 1: RLM-RACE for Identification of Internal miRNA Cleavage Sites on mRNA Targets

2.1.1. Reagents

1. Nuclease-free H_2O (DEPC-treated water).
2. Sodium acetate (3 M, pH 5.2).
3. Phenol:chloroform [1:1 (vol/vol)].
4. Ethanol (70 and 100%).
5. Isopropanol.
6. Glycogen (10 mg/ml).
7. dNTP mix: 10 mM each dATP, dCTP, dGTP, and dTTP.
8. DTT (0.1 M).

9. DNA loading buffer (5×).
10. DNA molecular markers (100 bp ladder).
11. Agarose gel.
12. Tris-acetate–EDTA buffer (1×).
13. Ethidium bromide (1 μg/ml).

2.1.2. Enzymes and Buffers

1. RNase inhibitor (40 U/μl).
2. RNase H (2 U/μl).
3. T4 RNA Ligase (New England BioLabs).
4. Ligation buffer (10×)*: 500 mM Tris–glycine, pH 7.8, 100 mM $MgCl_2$, 100 mM DTT, and 100 mM ATP.
5. Superscript III Reverse transcriptase (200 U/μl) (Invitrogen).
6. Reverse transcription buffer (10×)*: 200 mM Tris–glycine, pH 8.4, and 500 mM KCl.
7. Ampli-Taq Gold DNA polymerase (5 U/μl) (Applied Biosytems).
8. PCR Gold buffer (10×)*: 150 mM Tris–glycine, pH 8.0, and 500 mM KCl.
9. $MgCl_2$ (25 mM)*.

*All buffers are supplied by the manufacturers. Store enzymes and buffers as indicated by the suppliers.

2.1.3. RNA and DNA Purification Systems

1. A method for total RNA and poly(A)$^+$ RNA purification (10).
2. A method for purification of DNA from agarose gels (10).

2.1.4. Nucleic Acids and Oligonucleotides

1. RNA oligonucleotide adapter (see Note 1 for customize adapter design); for instance: 5′ CGACUGGAGCACGAG-GACACUGACAUGGACUGAAGGAGUAGAAA 3′(GeneRacer, Invitrogen).
2. ASPs (see Note 2 for primer design details); for instance: ASP-F 5′ CGACTGGAGCACGAGGACACTGA 3′ and nested-ASP-F 5′ GGACACTGACATGGACTGAAGGAGTA 3′ (GeneRacer, Invitrogen).
3. GSPs (see Note 2 for user-specific, primer design details).
4. Oligo(dT) primer. For example, 5′ GCTGTCAACGATACG CTACGTAACGGCATGACAGTG$(T)_{18}$ 3′ (GeneRacer, Invitrogen).

2.1.5. Vectors

1. A method to clone PCR products (see Note 3): either a linearized cloning vector or a quick cloning system such as the T/A cloning.

2.2. Protocol 2: RLM-RACE and Microarray Hybridization for High-Throughput Identification of miRNA Targets

Only the materials that have not been included in the previous section are listed.

2.2.1. Reagents

1. NaOH (1 M).
2. HCl (1 M).
3. rNTP mix: 75 mM each ATP, CTP, GTP, UTP.
4. GeneChip DNA Labeling Reagent 7.5 mM, Affymetrix.

2.2.2. Enzymes and Buffers

1. T7 RNA polymerase (20 U/μl). Recommended: MegaScript T7 RNA polymerase (Ambion).
2. Transcription buffer (10×)*: 400 mM Tris–glycine, pH 7.5, 100 mM NaCl, 60 mM $MgCl_2$, and 20 mM spermidine.
3. DNase I RNAse-free, 10 U/μl (GE Healthcare).
4. One-Phor-All buffer (10×)*: 1 M potassium acetate, 0.25 M Tris-acetate, pH 7.6, 0.1 M magnesium acetate, 5 mM β-mercaptoethanol, and 500 μg/mL BSA. (GE Healthcare).
5. Terminal deoxynucleotidyl transferase (30 U/μl) (Promega).
6. TdT buffer (5×)*: 500 mM cacodylate buffer, pH 6.8, 5 mM $CoCl_2$, and 0.5 mM DTT.

*All buffers are supplied by the manufacturers. Store enzymes and buffers as indicated by the suppliers.

2.2.3. DNA Purification System

1. A method for cleanup of DNA from enzymatic reactions (10).

2.2.4. Nucleic Acids and Oligonucleotides

1. T7-containing, ASP: 5′ *GCGTAATACGACTCACTATAG-GG*CGACTGGAGCACGAGGACACTGA (residues corresponding to the T7 promoter are underlined).
2. Oligo(dT) (see Subheading 2.1.4).
3. Random hexamers (0.5 μg/μl).
4. Affymetrix GeneChip DNA microarrays.

3. Methods

3.1. Protocol 1: RLM-RACE for Identification of Internal miRNA Cleavage Sites on mRNA Targets

Total or poly(A)+ RNA is directly ligated to the RNA adapter without any further enzymatic pretreatment. Ligated RNA is reverse transcribed using, preferentially, a GSP primer that anneals 300–400 nts downstream of the predicted cleavage site within the target mRNA. The 5′ end of a specific cleaved transcript is then obtained by PCR amplification of the first-strand cDNA using a reverse GSP and an ASP derived from the RNA adapter sequence. The 5′ RACE products are then gel-purified, cloned, and sequenced. The diagram in Fig. 1b summarizes all the reaction steps in the procedure.

3.1.1. Purification of High-Quality RNA

A number of methods for high-quality RNA purification are available (10). Intact RNA can be prepared using guanidine isothiocyanate/acid–phenol methods such as the TRIzol (Invitrogen) or TRI-Reagent (Ambion), although it is sometimes contaminated by polysaccharides that may inhibit reverse transcriptase PCR reactions. The method of choice for many investigators is a procedure that combines RNA extraction with denaturing reagents containing guanidine thiocyanate salts and selective binding of the RNA to silica-based membranes.

Poly(A)-selected RNA usually yields better results than total RNAs and it is highly recommended for amplification of difficult-to-amplify targets or targets from rare cleavage events. Poly(A)$^+$ RNA can be prepared from total RNA extractions by chromatography on oligo(dT)-cellulose columns that can be purchased from a number of commercial suppliers (see Note 4).

3.1.2. Ligation of the RNA Oligonucleotide Adapter to Untreated Cellular mRNA

1. Mix the RNA oligonucleotide (~250 ng) and the untreated poly(A)$^+$ RNA (50–150 ng) in a sterile microfuge tube. On average, the RNA oligonucleotide is used at a molar excess of 3–5 over target cellular RNA.
2. Incubate at 65°C for 5 min to relax any RNA secondary structure.
3. Cool on ice for 2 min and then spin-down briefly to collect the solution at the bottom of the tube.
4. Add the following components and mix by gently pipetting:

Ligation buffer* (10×)	1 μl
RNasin (40 U/μl)	1 μl
T4 RNA Ligase (5 U/μl)	1 μl
H_2O	to 10 μl
Final volume	10 μl

*1 μl of 10 mM ATP must be added to the reaction mix if not present in the ligation buffer.

5. Incubate at 37°C for 3 h.
6. Centrifuge briefly and place on ice.

3.1.3. Purification of the Ligation Products

The ligation product can be purified using spin-column chromatography where enzyme, salts, and non-ligated adapters are selectively removed from the reaction mix and the RNA is recovered

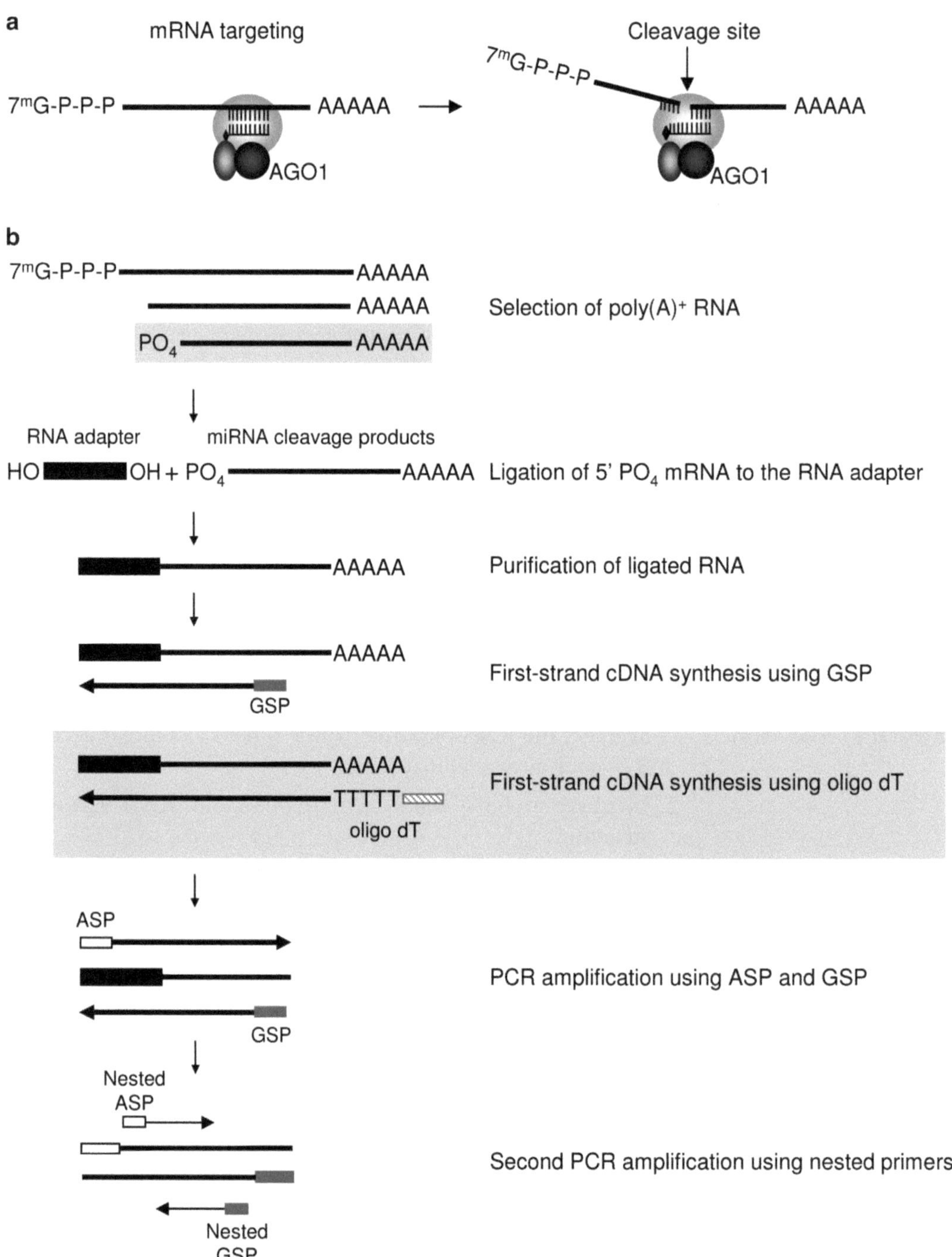

Fig. 1. Overview of the modified RLM-RACE procedure for precise mapping of miRNA-directed cleavage sites on target mRNAs. (**a**) miRNAs are recruited by AGO complexes to guide slicing of the target mRNA in the middle of the region of complementarity. (**b**) Experimental outline for RLM-RACE of untreated mRNAs.

enzyme-free. The Illustra MicroSpin S-300 Sephacryl columns (GE Healthcare) or equivalent commercial methods are suitable. Proceed as recommended by the manufacturer.

Alternatively, the ligated RNA can be extracted from the ligation reaction with phenol–chloroform, and precipitated with 0.1 volume of 3 M sodium acetate, pH 5.2, and 2.5 volume of absolute ethanol (10):

1. Add 90 μl of H_2O and 100 μl of phenol:chloroform to the ligation reaction and vortex vigorously for 30 s.
2. Centrifuge at maximum speed for 5 min at room temperature.
3. Transfer the aqueous top phase to a new eppendorf tube and add 2 μl of glycogen (10 mg/ml), 10 μl of 3 M sodium acetate, pH5.2. Mix well and then add 220 μl of ethanol. Vortex briefly and incubate on ice for 10–30 min.
4. Centrifuge at maximum speed for 20 min at 4°C.
5. Remove the supernatant with a pipet and add 500 μl of 70% ethanol.
6. Centrifuge at maximum speed for 5 min at 4°C and carefully remove the ethanol by pipetting.
7. Air-dry the pellet for 1–2 min at room temperature and resuspend in 10 μl of RNase-free water.

3.1.4. Reverse Transcription to Synthesize First-Strand cDNA

Next, the ligated RNA is selectively reverse transcribed into first-strand cDNA using an antisense GSP located ~200–400 nts downstream from the predicted miRNA cleavage site (see Note 5). The protocol below is optimized for reverse transcription with Superscript III RT (Invitrogen) (see Note 6). A control reaction setup without reverse transcriptase should be carried out in parallel.

1. In a microfuge tube, combine the following components:

Ligated RNA (from step Subheading 3.1.3)	8 μl
GSP (10 μM)	1 μl
dNTP mix (each at 10 mM)	1 μl

2. Incubate at 65°C for 5 min to denature RNA, chill on ice, and centrifuge briefly.

3. Add the following transcription reagents and mix gently by pipetting:

Reverse transcription buffer (10×)	2 μl
$MgCl_2$ (25 mM)	4 μl
DTT (0.1 M)	2 μl
RNasin (40 U/μl)	1 μl
Superscript III RT (200 U/μl)	1 μl
Final volume	20 μl

4. Incubate at 50–55°C for 1 h and then inactivate the reverse transcriptase by incubating at 85°C for 5 min.
5. Collect the reaction by brief centrifugation and chill on ice.
6. To degrade the RNA template, add 1 μl of RNase H (2 U) and incubate at 37°C for 20 min (see Note 7).
7. Centrifuge briefly. Use immediately for PCR amplification or store at –20°C.

3.1.5. PCR Amplification

cDNA amplification of 5′ ends is carried out using reverse GSP and forward ASP oligonucleotides. This protocol uses hot start and a touchdown scheme for PCR (see Note 8).

1. Set the following reaction in a 0.5-μl thin-wall PCR tube on ice:

PCR buffer (10×)	5 μl
$MgCl_2$ (25 mM)	3 μl
dNTP mix (each at 10 mM)	1 μl
GSP (10 μM)	1 μl
ASP (10 μM)	1 μl
RT template (see Note 9)	1–2 μl
H_2O	to 49.5 μl
Taq DNA polymerase (5 U/μl)	0.5 μl
Final volume	50 μl

2. Mix gently and, if necessary, overlay with mineral oil to prevent evaporation.

3. Transfer the tubes to a preheated thermocycler, and carry out 30–35 cycles of PCR according to the following typical cycling program (see Note 10).

Temperature (°C)	Time	Cycles
94	5 min	1
94	30 s	5
70	1 min/1 Kb	
94	30 s	5
65	1 min/1 Kb	
94	30 s	20–25
60	30 s	
72	1 min/1 Kb	
72	10 min	1

4. Higher yield of 5′ RACE products and reduced amplification of nonspecific products (observed as multiple bands or a smear) can be achieved if a second amplification round using nested GSP and ASP primers is performed (see Note 11). Setup the PCR reaction as indicated above using a dilution of the original PCR as a template (1.0–0.1%) and the GSP- and ASP-nested primers.
5. It is strongly recommended to include several 5′ RACE controls to facilitate interpretation of the results:
   (a) Omit the RT template to identify products that originate from amplification of contaminating genomic DNA.
   (b) Omit either the GSP or the ASP to identify products that result from nonspecific binding of the other primer in the pair.

3.1.6. Analysis of Gene-Specific 5′ RACE Products

1. After PCR, analyze an aliquot (5–10 μl) of the amplification reaction using 1–2% agarose gel electrophoresis in 1× TAE buffer and ethidium bromide staining (10). Include one well for molecular weight markers (100 bp DNA ladder).
2. Visualize bands on a UV transilluminator. Ultimately, the 5′ RACE procedure should produce a single band on the agarose gel representing all the mRNA products that result from cleavage events occurring at the region of complementarity between the miRNA and the target mRNA sequence (see Note 12). In case no bands are present in the samples, or if there is a complicated pattern of bands (i.e., a broad diffuse smear or multiple bands), perform a nested PCR as indicated above (see Note 13).

3.1.7. Cloning and Sequence Analysis of 5′ RACE Products

The specific PCR products must be gel-purified prior to cloning and sequencing. A number of standard methods for recovery of DNA from agarose gels are available and described in detail elsewhere (10).

For cloning PCR products, TA cloning vectors are recommended as they avoid using restriction enzymes. Ligation of the DNA insert into the appropriate cloning vector must be done according to the manufacturer's instructions. Once the PCR product is cloned, 10–20 clones should be selected for sequencing to ensure precise mapping of the 5′ ends representing all possible miRNA-directed cleavage sites.

3.2. Protocol 2: RLM-RACE and Microarray Hybridization for High-Throughput Identification of miRNA Targets

Poly(A)$^+$ RNA is ligated to the RNA oligonucleotide adapter and reverse transcribed from an oligo(dT) primer as indicated in the Subheadings 3.1.2 and 3.1.4, respectively. Synthesis of second-strand of cDNA is directed from an oligonucleotide primer derived from the RNA adapter that contains the sequence for the bacteriophage T7 promoter. The resulting double-stranded cDNA is subjected to in vitro transcription with T7 RNA polymerase, thus allowing lineal amplification of cDNA containing the ligated adapter. In vitro transcribed RNA is used as a template for synthesis of single-stranded cDNA, and the resulting cDNA is biotinylated and hybridized to Affymetrix DNA microarrays. A negative control consisting of a cDNA population in which the RNA adapter is not ligated to 5′ cleaved transcripts is critical in order to identify background hybridizations resulting from unspecific annealing of the RNA adapter and/or T7-oligonucleotide to RNA and cDNA.

It must be noted that although a certain degree of technical variation on microarray hybridizations is unavoidable, the major source of variation between different chips comes from the biological variability of samples. This problem can be overcome by pooling samples and by the analysis of at least three independent biological replicates (pools) and their corresponding negative controls.

Figure 2 summarizes all the reaction steps in the procedure.

3.2.1. Ligation of the RNA Oligonucleotide Adapter to Untreated Cellular mRNA

1. Mix the RNA oligonucleotide (~1 μg) and the untreated poly(A)$^+$ RNA (~500 ng) and proceed as described in Subheading 3.1.2 (see Note 14).
2. A negative control reaction setup without T4 RNA ligase must be carried out in parallel (see Note 15).

3.2.2. Purification of the Ligation Products

Proceed as described in Subheading 3.1.3. Excess of non-ligated RNA oligonucleotide adapter must be removed to prevent high background hybridization signals corresponding to nontarget genes in the microarray assays.

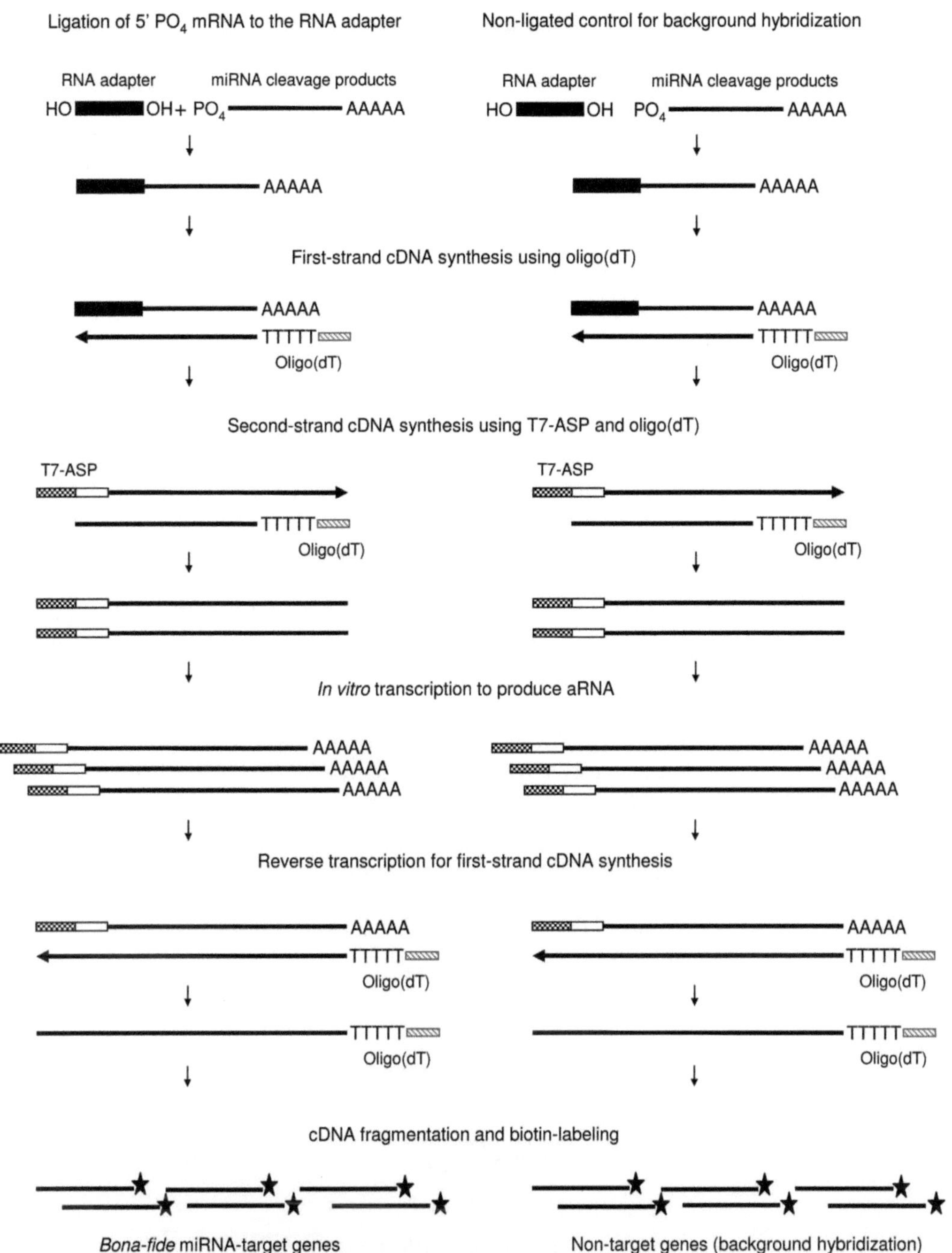

Fig. 2. Overview of the procedure for genome-wide identification of miRNA target mRNAs using selective RLM-RACE-based enrichment of 3′ end cleavage products, T7 RNA polymerase-mediated amplification of ligated products, and microarray hybridization. A negative control using non-ligated RNA provides a unique means to substract background hybridizations due to unspecific annealing of the RNA adapter and/or T7-ASP to RNA and cDNA.

3.2.3. Reverse Transcription to Synthesize First-Strand cDNA

Ligated RNA is reverse transcribed into first-strand cDNA using the Superscript III RT (Invitrogen) (see Note 16) and the oligo(dT) primer. A control reaction using the non-ligated RNA template must be necessarily included.

1. Setup reverse transcription reactions as described in Subheading 3.1.4.
2. Incubate at 50°C for 3 h. Chill on ice.
3. Add 10 μl of 1 M NaOH and incubate at 65°C for 30 min.
4. Collect the reaction by brief centrifugation and chill on ice.
5. Neutralize the reaction adding 10 μl of 1 M HCl and 10 μl of 3 M sodium acetate, pH 5.2 (see Note 17).
6. Immediately proceed with purification of first-strand cDNA.

3.2.4. Cleanup of First-Strand cDNA

Several methods for silica membrane-based purification of DNA from enzymatic reactions can be adopted for cleanup of single-stranded cDNA in the presence of a chaotropic agent. The MinElute columns (Qiagen) are suitable to eliminate the excess of dNTPs, primer, enzyme, and buffer, although other equivalent methods can also be used. Follow the specific instructions recommended by the manufacturer.

3.2.5. Double-Stranded cDNA Synthesis

Purified single-stranded cDNA serves as a template for second-strand cDNA synthesis using an ASP forward primer tailed at its 5′ end with the promoter sequence of a bacteriophage RNA polymerase (T7, T3 or SP6) (see Note 18). This protocol uses an oligonucleotide primer containing the T7 promoter sequence (T7-ASP) (see Subheading 2.2.4). Perform a negative control using cDNA from non-ligated RNA as indicated in Subheading 3.2.1. A negative control is critical for subtracting background hybridization in microarray assays, as cDNA obtained from non-ligated samples corresponds to nontarget genes.

1. Mix the following components in a 0.5-μl thin-wall PCR tube on ice:

PCR buffer (10×)	10 μl
$MgCl_2$ (25 mM)	6 μl
dNTP mix (each at 10 mM)	2 μl
T7-ASP (10 μM)	4 μl
RT template	60 μl
H_2O	to 99 μl
Taq DNA polymerase (5 U/μl)	1 μl
Final volume	100 μl

2. Transfer the tubes to a preheated thermal cycler and carry out the following program (see Note 19).

Temperature (°C)	Time (min)
94	5
58	1
72	10

3.2.6. Cleanup of Double-Stranded cDNA

Proceed as described in Subheading 3.2.4. Regardless of the method of choice for DNA cleanup, concentrate the sample to 8–10 μl using a concentrator/evaporator device.

3.2.7. In Vitro Transcription

In vitro transcription of T7 promoter-containing target cDNA products allows for enrichment and lineal amplification of cleavage mRNAs up to 50-fold relative to the starting material. The protocol below is optimized for in vitro transcription using the Megascript T7 kit (Ambion). If other systems or enzymes for in vitro transcription are used, follow manufacturer's instructions. Perform a negative in vitro transcription control using DNA from non-ligated RNA templates: amplified RNAs (aRNA) from the non-ligated RNA sample correspond to nontarget genes.

1. Unfreeze NTPs on ice and transcription buffer (10×) at room temperature.
2. Mix the following components in a microfuge tube at room temperature as transcription buffer (10×) contains spermidine that may precipitate at low temperature.

Template cDNA	8 μl
rNTP mix	8 μl (2 μl each of ATP, CTP, GTP and UTP)
Transcription buffer (10×)	2 μl
T7 RNA pol. enzyme mix	2 μl
Final volume	20 μl

3. Incubate at 37°C for 16 h.
4. Add 1 μl of DNase I (10 U/μl) to eliminate template DNA.
5. Inactivate DNase I by incubating for 15 min at 70°C.

3.2.8. Cleanup of aRNA

In vitro transcribed aRNA must then be purified from unincorporated NTPs, enzymes, and buffer components. A number of commercial kits designed specifically for the cleanup of transcription reactions (e.g., MEGAclear from Ambion or RNeasy from Qiagen) are suitable for this purpose. In this case, follow the specifications of the manufacturer. It is recommended to concentrate the aRNA sample to ~20 μl using a concentrator/evaporator device.

3.2.9. Reverse Transcription to Synthesize First-Strand cDNA from aRNA

The aRNA is subjected to reverse transcription using both oligo(dT) and random hexamers to increase final cDNA yield. Include a reverse transcription setup using aRNA template from the non-ligated sample control.

1. Prepare the following mixture in a microfuge tube:

aRNA	20 μl
Random Primers (0.5 μg/μl)	1.5 μl
Oligo(dT) (50 μM)	1.5 μl

2. Incubate at 70°C for 10 min, followed by incubation at 25°C for 10 min. Chill on ice and centrifuge briefly.
3. Add the following components to the reaction mix:

First-strand buffer (5×)	8 μl
DTT (0.1 M)	4 μl
dNTPs (each at 10 mM)	2 μl
RNasin (40 U/μl)	1 μl
Superscript III RT (200 U/μl)	3 μl
Final volume	40 μl

4. Incubate at 42°C for 2–3 h.
5. Collect the reaction by brief centrifugation and chill on ice.
6. Add 20 μl of 1 M NaOH and incubate at 65°C for 30 min to degrade template aRNA.
7. Neutralize the reaction adding 20 μl of 1 M HCl and 20 μl of 3 M sodium acetate, pH 5.2.
8. Immediately proceed with purification of first-strand cDNA as described in Subheading 3.2.4. (MinElute columns recommended).
9. Elute cDNA in 18 μl of sterile H_2O.

3.2.10. Fragmentation and Labeling of Single-Stranded cDNA

The target cDNA is then fragmented in the range of 50–200 bp and biotinylated at the 3′ termini using terminal deoxynucleotidyl transferase according to Affymetrix labeling protocols.

1. Prepare the following reaction mix (see Note 20):

One-Phor-All Buffer (10×)	2 µl
DNase I (1 U/µl)	1 µl
cDNA	17 µl
Total volume	20 µl

2. Incubate the reaction at 37°C for 10 min.
3. Inactivate DNase I at 98°C for 10 min.
4. Collect the reaction by brief centrifugation and chill on ice.
5. Combine the following reagents for terminal labeling:

Labeling reaction buffer (5×)	10 µl
GeneChip DNA labeling reagent (7.5 mM)	2 µl
Terminal deoxynucleotidyl transferase	2 µl
Fragmentation cDNA product	20 µl
H_2O	16 µl
Total volume	50 µl

6. Incubate the reaction at 37°C for 60 min.
7. Terminate the reaction by adding 2 µl of 0.5 M EDTA, pH 8.0.
8. The target cDNA is ready to be hybridized onto microarrays. Alternatively, it may be stored at −20°C for later use.

3.2.11. Microarray Hybridizations

The methodology described in this protocol is optimized for genome-wide identification of target-selected cDNA using Affymetrix DNA microarrays. We strongly recommend referring to the GeneChip Expression Analysis Technical Manual by Affymterix to learn on the hybridization details, including buffers and reagents required and washing and staining protocols. All the steps carried out in this protocol are identical to those indicated in the Affymetrix Technical Manual (Subheading 2, Chaps. 2 and 3; Eukaryotic Target Hybridization: Washing, Staining and Scanning) (see Note 21). Samples (RNA target-enriched and non-ligated RNA negative control) must be independently hybridized to microarrays. Target genes are identified as fold change increases in the experimental samples relative to negative controls.

3.2.12. Statistical Analysis of Microarrays for Identification of miRNA-Target Genes

The following methods are shown to give robust identification of miRNA-target genes and are used in this protocol: Robust multi-array average (11) for background correction and normalization, linear model methods (LiMMA) (12) for evaluation of differential gene expression, and the false discovery rate (FDR) method for correction of p-values in a multiple testing experiment (also known as BH) (13) (see Note 22). These methods can be run using different computational packages available at Bioconductor using the R software. The Bioconductor Package affylmGUI is a graphical user interface for Affymetrix analysis using the LiMMA microarray package which offers the possibility of performing the complete analysis of microarrays as indicated above (see Note 23).

Refer to the affylmGUI author's site for full documentation about this package. Here, it is a brief description of the steps in the process.

1. Create a file in tab-delimited text format containing the description of the CEL files according to the following format, and save as "my experiment.txt." This file must be saved in the same folder as CEL files.

Name	**FileName**	**Target**
miRNA target repl 1	Targets-1.cel	miRNA targets
miRNA target repl 2	Targets-2.cel	miRNA targets
miRNA target repl 3	Targets-3.cel	miRNA targets
Negative control repl 1	Negative-1.cel	Negative control
Negative control repl 2	Negative-2.cel	Negative control
Negative control repl 3	Negative-3.cel	Negative control

2. Load package affylmGUI in the R console.
3. The main window of affylmGUI appears, and from the File menu click on "New" to begin a new analysis.
4. Select Targets file: "my experiment.txt."
5. After the data have been loaded, select "Normalize" from the Normalization menu. Select RMA.
6. Once normalized, select "Linear Model Fit" from the Linear Model menu. The linear model is used to average data between replicate arrays and also look for variability between them.
7. Select "Compute Contrasts" from the Linear Model menu. This application performs the comparisons of interest between the arrays. Specify the contrast as follows: Contrast 1: miRNA targets minus Negative control. Select a name for the contrast.
8. From the Toptable menu, select "Table of Genes Ranked in order of Differential Expression." Select "All genes" to be displayed in the table, and then adjust p-values with "FDR." A full table corresponding to all the genes in the microarray is displayed, and can be copied and pasted into an Excel spreadsheet.
9. Filter genes over-represented in the miRNA-target experiments relative to negative controls. A good choice is to select

genes over-represented at least twice in the experiment (signal log ratio≥1, parameter M; and FDR≤0.05). This constitutes the candidate miRNA-target genes list.

4. Notes

1. The RNA oligonucleotide adapter must be approximately 40–45 bases in length and can be ordered as a synthetic RNA from a commercial source. Ideally, it must have adenines at the 3′ end to increase ligation efficiency and minimal secondary structure to provide a free 3′ end for efficient ligation. The RNA adapter sequence is specifically designed to lack the 5′-monophosphate to prevent either self-ligation or ligation to dephosporylated RNA. Alternatively, RNA oligonucleotides can be transcribed from a suitable plasmid vector linearized at a size about 70–100 bp downstream from a T7, T3, or SP6 RNA polymerase site.
2. All primers should be designed according to the following considerations (14):
   (a) Primers should be 20–24 bases in length with a ~50–70% GC content to allow for high annealing temperature and to improve the specificity of the PCR. The GC content at the 3′ end of the primer must be low to minimize extension by DNA polymerases at nontarget sites. Primers should have no secondary structure and no G as the 3′-terminal base. Primers should not self-hybridize or hybridize to the 3′ end of the other primers in the PCR reaction.
   (b) The annealing temperature (Tm of the primer) should be low enough to guarantee efficient annealing of the primer to the target, but high enough to prevent nonspecific binding. The optimal annealing temperature has to be determined experimentally and it should be similar to that for the primer derived from the RNA adapter sequence. Typically, an annealing temperature of 55–65°C can be a good starting point. The Tm can be estimated using 4× (G+C) + 2× (A+T), where G, C, A, or T represent the number of these bases in the primer sequence.
   (c) GSPs should be designed to anneal no closer than ~200 bases downstream of the predicted miRNA cleavage site within the target mRNA, so that it produces a resolvable product in PCR. If nested GSP are to be used, the spacing between the inner and the outer primers should be ~50–100 bases to produce PCR fragments that can be easily distinguishable in size.

3. Different types of plasmid vectors can be used for cloning amplification products using standard procedures (10). PCR cloning techniques using T/A overhangs are recommended to easily clone reaction products without restriction enzyme sites. T/A cloning systems exploit the non-template-dependent activity of *Taq* DNA polymerases that add a single deoxyadenosine to the 3′ end of the PCR products. As a result, PCR products can be directly ligated to linearized vectors that have single 3′ deoxythymidine residues. If a thermostable polymerase with extensive 3′ to 5′ exonuclease activity has been used for PCR amplification, 3′ A-overhangs can be added by incubation with Taq at the end of the cycling program. Alternatively, restriction sites can be incorporated into the GSP and ASP primers at their 5′ ends so the PCR fragments can be ligated into appropriately digested plasmid vectors.
4. The quality of the RNA is among the most critical factors influencing the outcome of the experiment. RNA integrity can be checked using a microfluidic chip-based automated electrophoresis system, such as the Experion system from Bio-Rad or the Bioanalyzer from Agilent. Alternatively, RNA can be assessed for integrity by traditional denaturing electrophoresis using 1% agarose and ethidium bromide staining. For total RNA, tight bands corresponding to the 28S (4.5 Kb) and 18S (1.9 Kb) ribosomal RNA should be obtained in a 2:1 ratio, while mRNA should run as a smear from 0.5 to 12 Kb. The purity of the RNA can be estimated by the A_{260}/A_{280} ratio (10).
5. Alternative to the use of a GSP for first-strand cDNA synthesis, an oligo(dT) primer can also be used to generate consistently a pool of non-gene-specific 5′ RACE products. This cDNA can then be subjected to a first round of amplification using the forward ASP and a nested antisense primer complementary to the oligo(dT) primer sequence. Gene-specific 5′ RACE PCR reactions are done with the nested ASP and GSPs. To prime the first-strand cDNA synthesis with an oligo(dT) primer, use 1 μl of oligo(dT) from a 50 μM stock. If random primers are used, preincubate the reaction mix at 25°C for 5 min for efficient binding of the random primers to the template.
6. Reverse transcriptases with reduced RNase H activity such as the cloned Avian Myeloblastosis Virus Reverse Transcriptase (AMV-RT) or the Superscript III (Invitrogen), a derivative of the Moloney Murine Leukemia Virus Reverse Transcriptase (M-MLV-RT), are highly recommended for first-strand cDNA synthesis. The reverse transcription reaction using cloned AMV-RT is typically performed at 45°C for 1 h. Optimize the reaction conditions according to the manufacturer's instructions when other RT enzymes are used.

7. After cDNA synthesis, thermal inactivation of the RT enzyme and RNase H treatment are recommended to prevent hairpin-primed second-strand synthesis and to destroy the template RNA in the cDNA-RNA hybrid.
8. Usually, "hot start" and touchdown PCR techniques are recommended to minimize nonspecific annealing and extension of primers prior to the initial denaturation step of the PCR process. Most of the commercial thermostable DNA polymerases that provide an automatic hot start are suitable.
9. The amount of RT template (typically less than 10% of the RT reaction) should be adjusted depending on the amount of starting RNA and the estimated abundance of the target mRNA cleavage product.
10. The optimal cycling parameters (annealing and extension times, and temperatures) must be determined empirically for each particular primer combinations. In general, use 60–65°C for annealing with primers with a Tm lower than 72°C. Lower temperatures facilitate the amplification of 5′ RACE products but also increase background amplification. The extension temperature for each DNA polymerase is indicated by the manufacturer. A general rule for extension is 1 min for each 1 kb of DNA. The number of cycles should be adjusted depending on the transcript abundance although it is not recommended more than 35 cycles as it increases the background. In general, failure to produce the expected bands indicates that the cycling conditions (or primer design) are inappropriate. All the parameters and conditions affecting the efficiency and specificity of the PCR are discussed elsewhere (10, 14). If an oligo(dT) primer was used to prime first-strand cDNA synthesis in reverse transcription reaction, the conditions used for the cDNA amplification with a nested oligo(dT) primer are the same as those for GSPs, with the exception that an extension time of 2.5 min is used.
11. A single amplification round may not be sufficient to generate enough specific products to be detectable by agarose gel electrophoresis and ethidium bromide staining. A second PCR using nested ASP and GSP primers is a convenient strategy for rare mRNA cleavage products. Nested ASP and GSP primers must anneal to an internal site within the cDNA sequence with respect to the primary ASP and GSP primers used for first-strand synthesis.
12. For first-time users, it is recommended to perform a positive 5′ RACE control using primers previously used for validated miRNA::mRNA target interactions. This helps to verify that all components and reactions in the protocol function properly. We typically carry out 5′ end amplification of cDNA

specific to the cleaved *SCL6-IV* (At4g00150) and *ARF10* (At2g28350) mRNAs using poly(A)-selected mRNA from Arabidopsis tissues. The SCL6-IV-1152R (5′ GGAGCCGA-GGAGGTCTAAGCTCAAA 3′) or ARF10-1675R (5′ CCGCT-TCCGCCTCTTCTTCCAAAA 3′) GSPs are used. The conditions used for reverse transcription and amplification are the same as those described in this protocol. In both cases, a unique gene-specific DNA fragment of the size predicted to be generated from a template resulting from a miRNA-guided cleavage event is detected by agarose electrophoresis and ethidium bromide staining (4, 15). Other mRNAs known to functionally interact with miRNAs might be a better choice as positive controls in other plant species.

13. A complete troubleshooting guide is supplied along with most of the commercial RLM-RACE kits (The Ambion's First-Choice RLM-RACE and the Invitrogen's GeneRacer kits are the most popular choices) used for validating miRNA target sites and can be consulted for advice if needed.
14. The use of poly(A)-selected RNAs is highly recommended in order to enhance the sensitivity of the detection of cleaved transcripts. Since there is no exponential PCR amplification of targets, the sensitivity of the method may be compromised if total RNA is used instead of poly$(A)^+$ RNA.
15. Genome-wide strategies involve the identification of differentially ligated mRNAs versus a suitable negative control in order to maximize the identification of *bona fide* targets above a background noise level.
16. The Superscript III RT (Invitrogen) is highly recommended as it allows longer incubation periods at high temperatures (42–55°C), thus permitting complete oligo(dT)-primed synthesis of first cDNA strand from cleaved mRNA templates.
17. After cDNA synthesis, removal of template RNA is absolutely needed to prevent unspecific priming during second-strand synthesis. RNase H treatment or alkaline hydrolysis at high temperatures can be used.
18. Tailing of cDNA molecules with an appropriate promoter sequence during second-strand synthesis is necessary to provide a means for efficient and specific amplification of cDNA products. Tailed cDNA can be then used as a template for an in vitro transcription in a cell-free reaction using appropriate RNA polymerases.
19. The annealing temperature is a critical parameter given that low temperatures may cause nonspecific priming, whereas high temperatures may prevent annealing. An annealing temperature of 55–60°C should be specific enough without compromising cDNA yield.

20. Dilute DNase I to 1 U/μl in 1× One-Phor-All Buffer. For cDNA fragmentation, it is recommended a DNase I treatment at 0.6–0.8 U/μg cDNA. The reaction here refers to a total cDNA content of ~1.5 μg, which is the typical yield after cDNA synthesis and purification.
21. A number of commercial microarrays for plant genomes are available. Alternative to single-labeling platforms (e.g., Affymetrix), two-color platforms can be used and require simultaneous and competitive hybridization of the target-enriched sample and the corresponding negative control, both labeled with different fluorescent dyes (typically Cy3/Cy5). Synthesis of single-stranded cDNA from aRNA templates, fluorochrome-based labeling and quantification of target samples largely depend on the system of choice and must be adapted according to the microarray platform used.
22. The identification of candidate miRNA-target genes is performed as in a classical gene expression experiment, in which hybridizations of target-enriched samples versus negative controls are compared. Several other statistical methods described for the analysis of microarrays at any of the following levels: background correction, normalization between arrays, statistical evaluation of differential gene expression and multiple testing adjustments are also suitable.
23. R is a free software environment for statistical computing and graphics, and Bioconductor is an open source software project for the analysis and comprehension of genomic data. The different programs and packages are freely downloadable from the following sites:
    (a) R: http://www.r-project.org/
    (b) Bioconductor: http://www.bioconductor.org/.
    (c) AffylmGUI: http://www.bioconductor.org/packages/release/bioc/html/affylmGUI.html
    (d) Author's site: http://bioinf.wehi.edu.au/affylmGUI/

Acknowledgments

This work was supported by grants to CL from the Spanish Ministry of Science and Technology (GEN2003-20222-CO2-00 and BIO2006-13107), and Comunidad de Madrid (Spain) (CCG07-CSIC/GEN-1804), and by grants to RS from the Ministry of Science and Technology (BIO2004-02502, BIO2007-66935, GEN2003-20218-C02-02 and CSD2007-00057-B). J.M.F.-Z. is supported by a contract I3P-Doctores (CSIC).

References

1. Eulalio, A., Huntzinger, E., and Izaurralde, E. (2008) Getting to the root of miRNA-mediated gene silencing *Cell* **132**, 9–14.
2. Voinnet, O. (2009) Origin, biogenesis, and activity of plant microRNAs *Cell* **136**, 669–87.
3. Kurihara, Y., and Watanabe, Y. (2004) Arabidopsis micro-RNA biogenesis through Dicer-like 1 protein functions *Proc Natl Acad Sci USA* **101**, 12753–8.
4. Llave, C., Xie, Z., Kasschau, K. D., and Carrington, J. C. (2002) Cleavage of Scarecrow-like mRNA targets directed by a class of Arabidopsis miRNA *Science* **297**, 2053–6.
5. Brodersen, P., Sakvarelidze-Achard, L., Bruun-Rasmussen, M., Dunoyer, P., Yamamoto, Y. Y., Sieburth, L., and Voinnet, O. (2008) Widespread translational inhibition by plant miRNAs and siRNAs *Science* **320**, 1185–90.
6. Brodersen, P., and Voinnet, O. (2009) Revisiting the principles of microRNA target recognition and mode of action *Nat Rev Mol Cell Biol* **10**, 141–8.
7. Scotto-Lavino, E., Du, G., and Frohman, M. A. (2006) Amplification of 5' end cDNA with 'new RACE' *Nat Protoc* **1**, 3056–61.
8. Scotto-Lavino, E., Du, G., and Frohman, M. A. (2006) 5' end cDNA amplification using classic RACE *Nat Protoc* **1**, 2555–62.
9. Franco-Zorrilla, J. M., Toro, F. J., Godoy, M., Pérez-Pérez, J., López-Vidriero, I., Oliveros, J. C., García-Casado, G., Llave, C., and Solano, R. (2009) Genome-wide identification of small RNA targets based on target enrichment and microarray hybridizations *Plant J* **59**, 840–50.
10. Sambrook, J., and Russell, D. W. (2001) *Molecular cloning: a laboratory manual*, Cold Spring Harbor Laboratory Press, New York.
11. Irizarry, R. A., Hobbs, B., Collin, F., Beazer-Barclay, Y. D., Antonellis, K. J., Scherf, U., and Speed, T. P. (2003) Exploration, normalization, and summaries of high density oligonucleotide array probe level data *Biostatistics* **4**, 249–64.
12. Smyth, G. K. (2004) Linear models and empirical Bayes methods for assessing differential expression in microarray experiments *Stat Appl Genet Mol Biol* **31**, Article 1.
13. Benjamani, Y., and Hochberg, Y. (1995) Controlling the false discovery rate *J R Stat Soc Ser C Appl Stat* **57**, 289–300.
14. Altshuler, M. L. (2006) *PCR troubleshooting: the essential guide*, Caister Academic Press.
15. Kasschau, K. D., Xie, Z., Allen, E., Llave, C., Chapman, E. J., Krizan, K. A., and Carrington, J. C. (2003) P1/HC-Pro, a viral suppressor of RNA silencing, interferes with Arabidopsis development and miRNA unction *Dev Cell* **4**, 205–17.

Chapter 15

A High-Throughput Sequencing-Based Methodology to Identify All Uncapped and Cleaved RNA Molecules in Eukaryotic Genomes

Matthew W. Endres, Rebecca T. Cook, and Brian D. Gregory

Abstract

How eukaryotic organisms regulate mRNA levels is a fundamental question in biology. It is clear that the steady-state concentration of RNA in a cell is determined by both the rate of its synthesis and turnover. Most of the early attention was focused on the study of gene transcription, while only recently posttranscriptional mechanisms have gained recognition for their regulatory importance. Posttranscriptional control of RNA levels is mediated by a number of pathways, including general RNA degradation and the more recently identified mechanism of RNA silencing (Belostotsky and Sieburth, Curr Opin Plant Biol 12:96–102, 2009; Garneau et al., Nat Rev Mol Cell Biol 8:113–126, 2007; Ramachandran and Chen, Trends Plant Sci 13:368–374, 2008; Xie and Qi, Biochim Biophys Acta 1779:720–724, 2008). Intriguingly, the regulatory RNA targets of both pathways can be identified by the distinguishing characteristic of a 5′ monophosphate. Specifically, removal of the 7-methyl guanosine cap attached to the 5′ end of mRNA molecules is an initiating signal for subsequent 5′–3′ RNA degradation. In RNA silencing, small RNA-directed, protein-mediated cleavage of an mRNA target generates a free 5′ monophosphate on the resulting 3′ RNA fragment (Belostotsky and Sieburth, Curr Opin Plant Biol 12:96–102, 2009; Garneau et al., Nat Rev Mol Cell Biol 8:113–126, 2007). Taking advantage of this chemical property (free 5′ monophosphate), a genome-wide approach for mapping all uncapped and cleaved transcripts in eukaryotic transcriptomes has been developed that we have termed "genome-wide mapping of uncapped and cleaved transcripts" (Gregory et al., Developmental Cell 14:854–866, 2008), which others have called degradome sequencing (Addo-Quaye et al., Curr Biol 18:758–762, 2008) or "parallel analysis of RNA ends" (German et al., Nat Biotechnol 26:941–946, 2008).

Key words: RNA degradation, RNA silencing, Genomics, High-throughput sequencing

1. Introduction

The biogenesis of a mature RNA transcript involves the complex interaction of many factors. For the purpose of this chapter, it will suffice to mention two steps in RNA maturation briefly: The additions of the 5′ cap and the 3′ poly(A) tail. Specifically, the processing

Tamas Dalmay (ed.), *MicroRNAs in Development: Methods and Protocols*, Methods in Molecular Biology, vol. 732, DOI 10.1007/978-1-61779-083-6_15, © Springer Science+Business Media, LLC 2011

of many precursor RNA molecules includes the addition of a guanine to its 5′ end that contains a methyl group at position 7 on its carbon ring. This structure is referred to as the 5′ cap. Meanwhile, at the 3′ end of the new RNA molecule adenosine (A) residues are added to form what is termed the poly(A) tail. These two structures serve to protect mature RNAs from destruction by exonucleolytic attack (8).

It is crucial that a cell eliminates aging and undesirable transcripts from its transcriptome, since these molecules can often be aberrant and potentially harmful (1, 2). To this end, the cell employs a complex, multifaceted surveillance system that selectively identifies RNA molecules for targeting to degradation pathways. In order for initiation of these RNA degradation mechanisms, one of the following three processes must occur: (1) removal of the 5′ cap followed by 5′→3′ exonucleolytic cleavage, (2) internal cleavage of the RNA molecule followed by 3′→5′ exonucleolytic degradation of the 5′ fragment and 5′→3′ exonucleolytic degradation of the 3′ fragment, or (3) deadenylation followed by 3′→5′ exonuclease activity (1, 2, 9–11). smRNA-directed, protein-mediated endonucleolytic cleavage utilized by RNA silencing pathways is one example of the second mechanism (3, 4, 6, 7, 11, 12). Therefore, RNA silencing is able to bypass the requirement for either deadenylation or decapping to destabilize regulatory target mRNAs by attacking the interior of these transcripts (see Fig. 1).

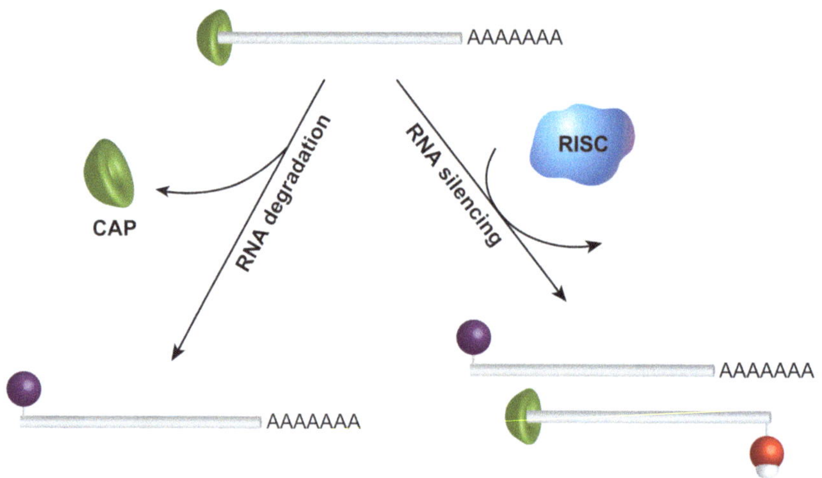

Fig. 1. Uncapped and proteolytically cleaved transcripts manifest the same chemical characteristic, a free 5′ monophosphate. On the left, an mRNA has been decapped to initiate its destruction by an RNA degradation pathway. The uncapped mRNA manifests a monophosphate (*purple ball*) on its 5′ end. On the right, smRNA-directed, protein-mediated cleavage of an mRNA target molecule by an RNA-induced silencing complex (RISC) produces a 3′ RNA fragment with a free 5′ monophosphate (*purple ball*). This protein-mediated RNA cleavage also produces a 5′ fragment with a free 3′ hydroxyl (oxygen is *red*, hydrogen is *white*).

To obtain a transcriptome-wide look at the importance of these processes in general RNA stability, we have developed a methodology that is described below. This methodology specifically interrogates mRNA molecules that are uncapped or proteolytically cleaved (see Fig. 1). To do this, we exploit the random amplification of complementary DNA (cDNA) ends (RACE) principle, but ignore the 5′ cap and selectively clone molecules with a 5′ monophosphate (5) (see Fig. 1). We can capture the free 5′ monophosphate on uncapped and cleaved RNA transcripts through ligation of an RNA adapter with a 3′ hydroxyl at its end. This modified RACE procedure is coupled with high-throughput sequencing technologies to get a genome-wide look at uncapped and cleaved RNAs. Using this approach, we not only define RNA degradation intermediates, but can also identify all regulatory targets that are endonucleolytically cleaved by smRNA-directed protein complexes genome-wide (see Fig. 1).

2. Materials

2.1. Purification of Ribosomal RNA-Depleted PolyA+ RNA

1. Oligotex Direct mRNA Mini Kit (Qiagen, Valencia, CA).
2. Ribominus™ Plant Kit for RNA-seq (Invitrogen, Carlsbad, CA).
3. 100% Ethanol. Store at room temperature.
4. 3 M Sodium acetate (NaOAc), pH = 5.2–5.5. This solution should be made in diethylpyrocarbonate (DEPC)-treated water (DEPC-d_2H_2O) (see Note 1). Store at room temperature.
5. 80% Ethanol: 100% ethanol is diluted to 80% using DEPC-d_2H_2O. Store at room temperature.

2.2. Ligation of 5′ RNA Adapter

1. T4 RNA ligase (10 U/μL) (Promega, Madison, WI). The accompanying buffer will be used in the ligation step.
2. RNaseOUT Recombinant Ribonuclease Inhibitor (40 U/μL) (Invitrogen, Carlsbad, CA).
3. 10 mM ATP: 100 mM ATP (Roche, Indianapolis, IN) is diluted to 10 mM in DEPC-treated water (see Note 1) and immediately frozen in single use (20 μL) aliquots at −20°C.
4. 5′ RNA adapter (5 μM): The adapter needs to possess hydroxyl groups at both the 5′ and 3′ ends to control against (1) the RNA substrates ligating to the wrong end of the adapter, as well as (2) adapter–adapter ligation products. For our experiments, we utilize the Illumina© Genetic Analyzer II sequencing technology. Therefore, our 5′ RNA adapter sequence is 5′-GUUCAGAGUUCUACAGUCCGACGAUC-3′, which

is called the Illumina 5′ small RNA (smRNA) sequencing adapter.

5. 0.2 mL PCR tubes in strips of eight with caps (USA Scientific, Ocala, FL). These specific PCR strip tubes are used because they come certified DNase, RNase, and pyrogen-free.
6. Phenol:chloroform:isoamyl alcohol, 25:24:1, pH 7.9 (Ambion, Austin, TX). This solution is light-sensitive, so it should be split into 20-mL aliquots and stored at 4°C in polypropylene tubes covered with an aluminum foil. For long-term storage, the 20-mL aliquots in aluminum foil-covered polypropylene tubes should be kept at –20°C.

2.3. Double-Stranded Complementary DNA Synthesis

1. Superscript first-strand synthesis system for RT-PCR (Invitrogen, Carlsbad, CA). This system contains the Oligo(dT)$_{12-18}$ (0.5 μg/μL), 10× RT buffer, 25 mM magnesium chloride ($MgCl_2$), 100 mM dithiothreitol (DTT), RNaseOUT Recombinant Ribonuclease Inhibitor (40 U/μL), SuperScript II reverse transcriptase (RT) (50 U/μL), and *Escherichia coli* RNase H (2 U/μL) that are used during the first-strand cDNA synthesis step of this protocol.
2. Nuclease-free water (Ambion, Austin, TX).
3. 12.5 mM dNTPs. To begin, 100 mM dNTP (dATP, dCTP, dGTP, and dTTP) stocks are diluted to 12.5 mM using nuclease-free water.
4. PCR primer specific for the 5′ RNA adapter. For our purposes, we are using the Illumina Genetic Analyzer II technology, so we utilize a 5′ PCR primer with the sequence, 5′-AATGATACGGCGACCACCGACAGGTTCAGAG TTCTACAGTCCGA-3′. However, different primers can and should be utilized depending on the selected high-throughput sequencing technology.
5. Phusion high-fidelity DNA Polymerase (2 U/μL) (New England Biolabs, Ipswich, MA) (see Note 2).
6. QIAquick PCR Purification Kit (Qiagen, Valencia, CA).

2.4. Shearing of Double-Stranded cDNA

1. Bioruptor (Diagenode, Sparta, NJ) (see Note 3).
2. Buffer EB (Qiagen, Valencia, CA).
3. Glycogen (5 mg/mL) (Ambion, Austin, TX).
4. Nuclease-free water (Ambion, Austin, TX).

2.5. Producing Adapter-Ligated High-Throughput Sequencing Libraries

1. Genomic DNA Sample Prep Kit (Illumina, La Jolla, CA) (see Note 4).
2. QIAquick Gel Extraction Kit (Qiagen, Valencia, CA).
3. Nuclease-free water (Ambion, Austin, TX).

2.6. Amplification of Adapter-Ligated Sequencing Libraries

1. Nuclease-free water (Ambion, Austin, TX).
2. Phusion high-fidelity DNA polymerase mastermix with HF buffer (New England Biolabs, Ipswich, MA) (see Note 2).
3. 5′ DNA primer specific for the ligated RNA adapter (5′ RNA adapter primer) (5′-AATGATACGGCGACCACCGACAGG TTCAGAGTTCTACAGTCCGA-3′, see Note 5).
4. 3′ DNA primer specific for the ligated 3′ genomic DNA (gDNA) sequencing adapter (3′ DNA adapter primer) (5′-CAAGCAGAAGACGGCATACGAGCTCTTC CGATCT-3′, see Note 5).
5. QIAquick PCR Purification Kit (Qiagen, Valencia, CA).

2.7. Sequencing Library Validation

1. 6% TBE gels, 1.0 mm, 10 wells (Invitrogen, Carlsbad, CA).
2. TBE gel running buffer: Prepare 10× stock with 108 g Tris base, 55 g boracic acid (boric acid), and 40 mL 0.5 M ethylenediamine tetraacetic acid (pH 8.0) in 1 L of water (see Note 1). Dilute 100 mL with 900 mL water (see Note 1) for use.
3. 10 mg/mL ethidium bromide (EtBr): Prepare stock by mixing 500 mg of EtBr powder with 50 mL water (see Note 1). Use 100 μL per 2 L of 1× TBE gel running buffer.
4. NanoDrop 2000 (Thermo Scientific, Wilmington, DE).

3. Methods

Many common techniques for defining the 5′ ends of mRNA, such as 5′ RACE, often use the cap as a molecular marker. The presence or absence of a 5′ cap moiety can be utilized to distinguish between functional mRNAs (capped) and those that have been uncapped or proteolytically cleaved. More specifically, those RNA molecules that are uncapped or proteolytically cleaved have an exposed 5′ monophosphate. Therefore, we describe below the steps that are necessary to take advantage of this chemical property to obtain a genome-wide view of RNA degradation intermediates (Fig. 2).

Briefly, the process begins by selection of ribosomal RNA (rRNA)-depleted polyA$^+$ mRNA to ensure that the initial uncapped and proteolytically cleaved mRNA intermediates are interrogated by the described methodology. Next, RNA degradation intermediates are selected by ligation of an RNA adapter to the free 5′ monophosphate on these molecules. The adapter-ligated RNA degradation intermediates are converted into double-stranded cDNA (dscDNA) and sheared to molecules that are

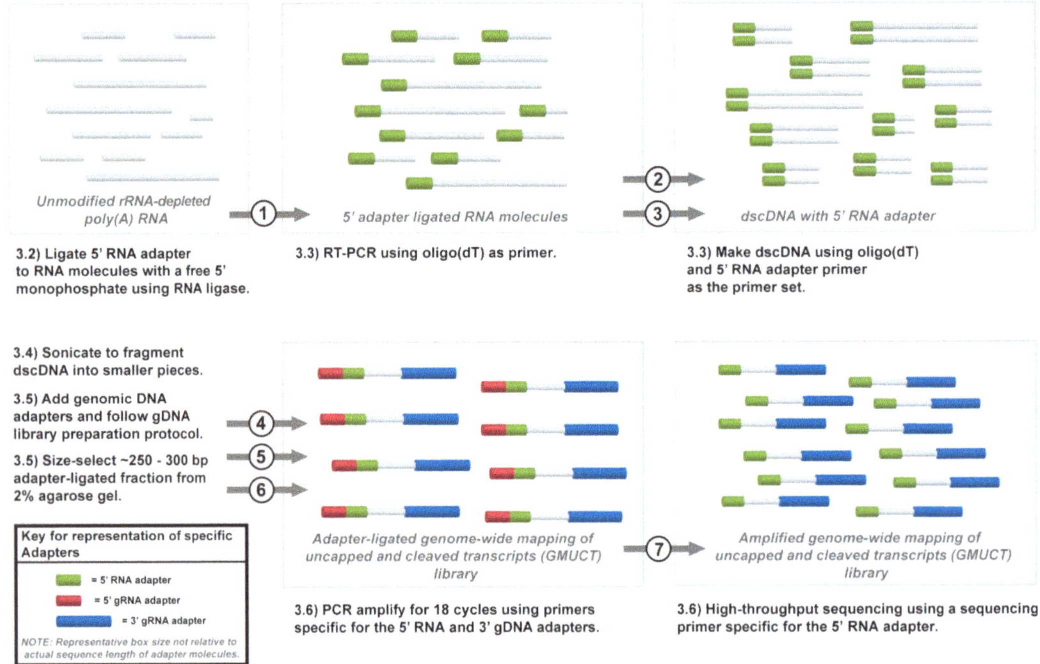

Fig. 2. A schematic of genome-wide mapping of uncapped and cleaved transcripts (GMUCT) library preparation. Within this chapter is a detailed, step-by-step explanation of the procedure for producing these high-throughput sequencing libraries.

properly sized for high-throughput sequencing library construction. To obtain a genome-wide view of all uncapped and proteolytically cleaved RNA molecules, the sheared dscDNA is utilized as a substrate in high-throughput DNA sequencing library preparation (Fig. 2).

3.1. Purification of rRNA-Depleted PolyA⁺ RNA

1. 50 μg of total RNA (see Note 6) from the organism, tissue, and genotype(s) of choice is column purified using the Oligotex mRNA kit following the included purification protocol.
2. During the elution of polyA$^+$ RNA from the Oligotex resin, it is suggested to use the highest recommended volume (100 μL) of hot (70°C) buffer OEB. Upon pipetting the 100 μL of hot (70°C) buffer OEB onto the column containing the Oligotex–RNA complexes, make sure to pipette the resin up and down thoroughly (~7×) to ensure complete resuspension. The elution step is repeated an additional time. Therefore, the purified polyA$^+$ RNA is contained at the end of this selection protocol in 200 μL of buffer OEB.
3. The 200 μL volume following selection of polyA$^+$ RNA in step 1 is 10× greater than the volume that is used in the Ribominus protocol (20 μL). Therefore, ethanol precipitation is required before proceeding to rRNA depletion of the polyA$^+$ RNA.

4. To do this, the 200 μL of buffer OEB containing the polyA$^+$ RNA is moved to a new, RNase-free 1.5-mL microcentrifuge tube and mixed with 20 μL 3 M NaOAc (pH = 5.2–5.5) and 600 μL of 100% ethanol.
5. This ~820 μL solution is mixed vigorously to ensure integration of the components, and frozen at −80°C for at least an hour (see Note 7).
6. After an hour at −80°C, the 1.5-mL microcentrifuge tube containing the precipitation reaction is centrifuged at maximum speed (14K rpm (20,817 *g*)) for 60 min at 4°C.
7. Following centrifugation, the supernatant is discarded and the polyA$^+$ RNA pellet is washed with 750 μL of 80% ethanol. The microcentrifuge tube containing the polyA$^+$ RNA and the 80% ethanol is centrifuged at maximum speed (14K rpm (20,817 *g*)) for 5 min at 4°C to ensure that the RNA pellet remains attached to the side of the tube.
8. Following centrifugation, the 80% ethanol is discarded and the microcentrifuge tube containing the polyA$^+$ RNA pellet is placed upside down on a paper towel for 10–15 min to dry.
9. After the residual 80% ethanol has evaporated, the polyA$^+$ RNA pellet is resuspended in 20 μL DEPC-d_2H_2O at room temperature for 10–15 min.
10. The 20 μL of polyA$^+$ RNA is then depleted for 18S and 28S rRNA molecules utilizing Ribominus reactions, following the included purification protocol. This procedure makes use of six plant-specific biotinylated LNA oligonucleotide rRNA probes that are supplied by the manufacturer. After this purification, the rRNA-depleted polyA$^+$ RNA is in a volume of ~500 μL. Therefore, before proceeding with 5′ adapter ligation, the rRNA-depleted polyA$^+$ RNA must be precipitated and resuspended in a lower volume.
11. To do this, the ~500 μL solution containing the rRNA-depleted polyA$^+$ RNA is moved to a new, RNase-free 2.0-mL microcentrifuge tube and mixed with 50 μL 3 M NaOAc (pH 5.2–5.5) and 1,500 μL of 100% ethanol.
12. This ~2 mL solution is mixed vigorously to ensure integration of the components, and frozen at −80°C for at least an hour (see Note 7).
13. After an hour at −80°C, the 2.0-mL microcentrifuge tube containing the precipitation reaction is centrifuged at maximum speed (14K rpm (20,817 *g*)) for 60 min at 4°C.
14. Following centrifugation, the supernatant is discarded and the rRNA-depleted polyA$^+$ RNA pellet is washed with 750 μL of 80% ethanol. The microcentrifuge tube containing the rRNA-depleted polyA$^+$ RNA and the 80% ethanol is

centrifuged at maximum speed (14K rpm (20,817 *g*)) for 5 min at 4°C to ensure that the RNA pellet remains attached to the side of the tube.

15. Following centrifugation, the 80% ethanol is discarded and the microcentrifuge tube containing the rRNA-depleted polyA+ RNA pellet is placed upside down on a paper towel for 10–15 min to dry.
16. After the residual 80% ethanol has evaporated, the rRNA-depleted polyA+ RNA pellet is resuspended in 4.7 μL DEPC-d_2H_2O at room temperature for 10–15 min.

3.2. Ligation of 5′ RNA Adapter

1. The 4.7 μL of rRNA-depleted polyA+ RNA is moved to a new, RNase-free 0.2-mL PCR strip tube (eight tubes per strip) and mixed with 1.3 μL of 5 μM 5′ RNA adapter.
2. The rRNA-depleted polyA+ RNA and adapter mix (RNA/adapter mix) is then heated to 65°C for 5 min in a thermocycler with a heated lid (see Note 8).
3. After 5 min at 65°C, the RNA/adapter mix is cooled to 4°C for 2 min in the thermocycler. This heating and subsequent cold treatment locks the RNA molecules into a denatured confirmation, which is necessary for efficient 5′ adapter ligation.
4. Following 2 min at 4°C, the RNA/adapter solution is mixed with 1 μL 10× RNA ligase buffer (included with enzyme), 1 μL 10 mM ATP, 1 μL RNaseOUT, and 1 μL T4 RNA ligase. This reaction is then incubated in a thermocycler (with its heated lid off) for 1 h at 37°C, followed by 5 h at 20°C, and completed at 4°C overnight (see Note 8).
5. Following the overnight incubation, the 5′ adapter ligation reaction was cleaned up with a phenol:chloroform extraction. Specifically, 90 μL of DEPC-d_2H_2O is added to the ligation reaction bringing the volume to 100 μL total.
6. The ligation reaction is then moved to a new, RNase-free 1.5-mL microcentrifuge tube and mixed vigorously with 100 μL of phenol:chloroform:IAA (25:24:1) by vortexing for 30 s.
7. The solution is then centrifuged at maximum speed (14K rpm (20,817 *g*)) for 5 min at 4°C.
8. The aqueous (top) phase is then transferred to a new, RNase-free 1.5-mL microcentrifuge tube (~100 μL) and mixed with 10 μL 3 M NaOAc (pH = 5.2–5.5) and 300 μL of 100% ethanol.
9. This ~410 μL solution is mixed vigorously to ensure integration of the components, and frozen at −80°C for at least an hour (see Note 7).

10. After an hour at −80°C, the 1.5-mL microcentrifuge tube containing the precipitation reaction is centrifuged at maximum speed (14K rpm (20,817 *g*)) for 60 min at 4°C.
11. Following centrifugation, the supernatant is discarded and the adapter-ligated RNA pellet is washed with 750 μL of 80% ethanol. The microcentrifuge tube containing the adapter-ligated RNA and the 80% ethanol is centrifuged at maximum speed (14K rpm (20,817 *g*)) for 5 min at 4°C to ensure that the RNA pellet remains attached to the side of the tube.
12. Following centrifugation, the 80% ethanol is discarded and the microcentrifuge tube containing the adapter-ligated RNA pellet is placed upside down on a paper towel for 10–15 min to dry.
13. After the residual 80% ethanol has evaporated, the adapter-ligated RNA pellet is resuspended in 8 μL DEPC-d_2H_2O at room temperature for 10–15 min.

3.3. dscDNA Synthesis

1. The adapter-ligated RNA is then used in an Oligo(dT)-primed cDNA synthesis using SuperScript II reverse transcriptase (RT). Specifically, the 8 μL of adapter-ligated RNA is moved to a new, RNase-free 0.2-mL PCR strip tube (eight tubes per strip), and mixed with 1 μL of Oligo$(dT)_{12-18}$ primer and 1 μL of 12.5 mM dNTPs.
2. This mix is then heated to 65°C for 5 min in a thermocycler with a heated lid (see Note 8).
3. After 5 min at 65°C, the mix is cooled to 4°C for 2 min in the thermocycler. This heating and subsequent cold treatment locks the RNA molecules into a denatured confirmation, which is necessary for efficient first-strand cDNA synthesis.
4. Following 2 min at 4°C, this 10 μL solution is mixed with 2 μL 10× RT buffer, 4 μL 25 mM $MgCl_2$, 2 μL 100 mM DTT, 1 μL RNaseOUT, and 1 μL Superscript II RT. This 20-μL reaction is then incubated in a thermocycler with a heated lid at 42°C for 1 h, followed by 15 min at 70°C, and completed at 4°C.
5. Once the cDNA synthesis reaction reaches 4°C, 1 μL of *E. coli* RNase H (2 U) is added to the reaction and this mix is incubated at 37°C for 20 min to digest all RNA molecules.
6. The RNA-free cDNA is then subjected to second-strand cDNA synthesis. To do this, the 21 μL first-strand cDNA synthesis reaction is mixed with 17.5 μL nuclease-free water, 10 μL 5× Phusion HF Buffer, 1 μL 12.5 mM dNTPs, 0.5 μL 5′ RNA adapter-specific primer, 0.5 μL Oligo$(dT)_{12-18}$ primer, and 0.5 μL Phusion high-fidelity DNA polymerase. This second-strand cDNA synthesis is subjected to the following cycle in a thermocycler with a heated lid (see Note 8).

Second-strand cDNA synthesis cycle:

(a) 94°C for 2 min
(b) 94°C for 30 s
(c) 72°C for 2 min 30 s
(d) Go to step 2, four additional times
(e) 94°C for 30 s
(f) 70°C for 2 min 30 s
(g) Go to step 5, four additional times
(h) 94°C for 30 s
(i) 60°C for 30 s
(j) 72°C for 2 min 30 s
(k) Go to step 8, 11 additional times
(l) 72°C for 10 min
(m) Store at 4°C.

7. Following second-strand cDNA synthesis, the reaction is cleaned up using the QIAquick PCR Purification Kit following the included protocol. The purified dscDNA is eluted from the purification column in 50 μL of Buffer EB.

3.4. Shearing of dscDNA

1. To increase the dscDNA volume to the required level of 300 μL for shearing in a Bioruptor, an additional 250 μL of Buffer EB is added to the corresponding 1.5 mL microcentrifuge tube.
2. The 300 μL sample is placed into a Bioruptor, making sure that the level of the dscDNA-containing liquid in the 1.5-mL microcentrifuge tube is below the surface of water in the Bioruptor reservoir. Additionally, the Bioruptor reservoir needs to be half-full of ice.
3. Program the Bioruptor with the following settings: 30 s on and 2 min off for six treatments consecutively (15 min total). Then turn the Bioruptor on to the longest setting (15 min). The samples will be done shearing after 13 min total since the last 2 min of the 15-min cycle is the off setting. Therefore, to speed up this part of the protocol, the dscDNA samples can be moved to step 4 of this protocol after 13 min in the Bioruptor.
4. Upon completion of cycle, the dscDNA sample is centrifuged to collect all liquid at the bottom of the sample tube, and then replaced into Bioruptor, making sure that all of the dscDNA-containing liquid is below the surface of water in the Bioruptor reservoir. Also, between cycles make sure to replenish any ice that has melted in the Bioruptor reservoir by replacing some water volume with ice.

5. This procedure is repeated two more times for a total of three cycles (see Note 9).
6. Following the third shearing cycle, the 300 μL dscDNA sample is mixed with 30 μL 3 M NaOAc (pH = 5.2–5.5), 3 μL glycogen (5 mg/mL) and 900 μL of 100% ethanol.
7. This ~1,233 μL solution is mixed vigorously to ensure integration of the components, and frozen at −80°C for at least an hour (see Note 7).
8. After an hour at −80°C, the 1.5-mL microcentrifuge tube containing the precipitation reaction is centrifuged at maximum speed (14K rpm (20,817 *g*)) for 60 min at 4°C.
9. Following centrifugation, the supernatant is discarded and the sheared dscDNA pellet is washed with 750 μL of 80% ethanol. The microcentrifuge tube containing the sheared dscDNA and the 80% ethanol is centrifuged at maximum speed (14K rpm (20,817 *g*)) for 5 min at 4°C to ensure that the dscDNA pellet remains attached to the side of the tube.
10. Following centrifugation, the 80% ethanol is discarded and the microcentrifuge tube containing the sheared dscDNA pellet is placed upside down on a paper towel for 10–15 min to dry.
11. After the residual 80% ethanol has evaporated, the sheared dscDNA is resuspended in 30 μL nuclease-free water at room temperature for 10–15 min.

3.5. Producing Adapter-Ligated High-Throughput Sequencing Libraries

For high-throughput sequencing library preparation, a Genomic DNA Sample Preparation Kit will be utilized (see Note 10). Therefore, sequencing library preparation is performed as described in the protocol that accompanies the sample preparation kit of choice. The following is a brief synopsis (with a couple slight modifications) of the steps involved in the construction of GMUCT sequencing libraries with the Illumina Genomic DNA Sample Prep Kit (see Note 10).

1. The first step is to repair the sheared ends of the dscDNA molecules, and is performed exactly as described in the accompanying library preparation protocol.
2. The second step is the addition of an adenine (A) base to the 3′ ends of the dscDNA molecules, and is performed exactly as described in the accompanying library preparation protocol.
3. The third step is the ligation of the DNA adapters to the ends of the dscDNA fragments. We have added a slight modification to the protocol for this procedure. Specifically, the adapter oligonucleotide mix is diluted 1:10 using nuclease-free water and then used at the volume specified in the protocol (for Illumina 10 μL).

4. The final step where the manufacturer's protocol is employed corresponds to the gel purification of ligation products procedure. We have also slightly modified this protocol. Specifically, a gel slice corresponding to 250–300 nucleotides should be excised and subsequently purified using the QIAquick Gel Extraction Kit following the included protocol. The adapter-ligated products are eluted from the purification column in 30 μL of Buffer EB. Upon obtaining 30 μL of gel-purified 250–300 nucleotide adapter-ligated molecules, it is time to proceed to the amplification protocol described below in Subheading 3.6.

3.6. Amplification of Adapter-Ligated Sequencing Libraries

1. To amplify the GMUCT sequencing library, 15 μL of gel-purified adapter-ligated dscDNA is moved to a new, RNase-free 0.2-mL PCR strip tube (eight tubes per strip) and mixed with 9 μL nuclease-free water, 25 μL Phusion high-fidelity DNA polymerase mastermix with HF buffer, 0.5 μL 5′ RNA adapter primer, and 0.5 μL 3′ DNA adapter primer.
2. The PCR is then placed into a thermocycler with a heated lid (see Note 8) and amplified utilizing the following cycle.

   GMUCT library amplification cycle:

   (a) 98°C for 30 s

   (b) 98°C for 10 s

   (c) 60°C for 30 s

   (d) 72°C for 30 s

   (e) Go to step 2, 17 additional times

   (f) 72°C for 10 min

   (g) Store at 4°C.
3. Upon completion of the thermocycler program, the amplified GMUCT high-throughput sequencing library is purified using the QIAquick PCR Purification Kit following the manufacturer's protocol. The one modification to the protocol is that the elution of the purified library products is performed by adding 30 μL of buffer EB to the purification column, allowing this to incubate for 5 min, and then spinning the liquid through for collection in a nuclease-free 1.5-mL microcentrifuge tube.

3.7. Sequencing Library Validation

1. Determine the concentration of the resulting GMUCT library by measuring its absorbance at 260 nm. The yield from the protocol should be ~0.75–1.50 μg of library DNA when starting with 50 μg of total RNA.
2. Measure the 260/280 ratio. It should be ~1.8 (see Note 11).

3. Load 5 μL of the library on a 6% TBE acrylamide gel and check that the size range is as expected (should be in the size range 215–265 nucleotides (nts)) (see Note 12).
4. To determine the molar concentration of the library, examine the gel image and estimate the median size of the library smear. This is generally about 240 nts. Multiply this size by 650 (the approximate molecular mass of a base-pair) to get the molecular weight of the library. Use this number to calculate the molar concentration of the library.
5. See Note 13.

4. Notes

1. Unless stated otherwise, all solutions should be prepared in water that has a resistivity of 18.2 MΩ cm. For all RNA work, water with a resistivity of 18.2 MΩ cm should be treated with DEPC (Fisher, Pittsburgh, PA) to assure inactivation of contaminating RNases. This standard is referred to as "water" in this text.
2. For this protocol, the 5× HF buffer is utilized in the synthesis of the dscDNA as described in Subheading 3.3, as well as GMUCT library amplification (Subheading 3.6). It is worth noting that any other high-fidelity DNA polymerase can be utilized during these steps of the protocol.
3. Other disruption devices including sonicators or nebulizers can also be used for this step. However, the protocol described herein was optimized for the specific use of a Bioruptor for dscDNA shearing.
4. This kit contains the reagents necessary to construct a high-throughput sequencing library utilizing dscDNA as the substrate. It is of note that if another high-throughput sequencing platform is going to be utilized, then that company's corresponding gDNA sequencing library construction kit should be employed.
5. Regardless of their source, it is of the utmost importance that the amplification primers be purified by HPLC or some other methodology to ensure that they are truly a monoculture of DNA molecules with precisely the same sequence.
6. We have been successful in constructing complex, well-distributed GMUCT libraries from starting quantities of RNA as low as 15 μg. However, we have not been successful (as of yet) with quantities below this lower threshold. For those trying this methodology for the first time, it is recommended

to begin with as much starting total RNA as possible to ensure that the final GMUCT library will be complex and well-distributed across the entire transcriptome (not just highly expressed transcripts).

7. One can proceed after an hour at –80°C. However, this is a good place in the protocol to stop if a break in library construction is required.
8. The best results in GMUCT library construction are obtained when this step is performed in a thermocycler.
9. This procedure can be modulated to obtain properly sized substrates for next-generation sequencing library preparation using gDNA or dscDNA from numerous organisms as starting material. Thus far, we have utilized the described protocol for shearing *Arabidopsis thaliana*, human, and *Drosophila melanogaster* dscDNA and gDNA.
10. For preparation of our GMUCT sequencing libraries, we have utilized the Genomic DNA Sample Prep Kit from Illumina (Illumina, La Jolla, CA). Therefore, the steps involved in library preparation will be described as specified by the protocol that accompanies this sample preparation kit.
11. We have obtained libraries with 260/280 ratios around 1.7 that produced very good sequence. Therefore, a 260/280 ratio of >1.7 will more than suffice for most GMUCT sequencing libraries.
12. This is the expected size range because the amplification is done with primers specific for the 5′ RNA adapter and the 3′ DNA adapter. Therefore, the 5′ DNA adapter (~35 nts) that is attached as part of the procedure in Subheading 3.5 is lost during the amplification cycle detailed in Subheading 3.6.
13. The first time this protocol is utilized, 2 μL of the library should be cloned into a sequencing vector (e.g. TOPO Blunt Cloning (Invitrogen, La Jolla, CA)). Then sequence at least ten individual clones by conventional Sanger sequencing. Verify that the adapter sequences are correct, and that the insert sequences correspond to a transcript encoded in the genome of the interrogated organism. Furthermore, confirm that the strand information of that mRNA is maintained. It is important to note that this protocol is strand-specific in nature.

Acknowledgments

The authors thank Drs. Joseph R. Ecker, Ronan C. O'Malley, and Ryan Lister for their help and support in the designing of this methodology. Work in the Gregory lab is supported by the

University of Pennsylvania, and in part by the Penn Genome Frontiers Institute and a grant with the Pennsylvania Department of Health. The Department of Health specifically disclaims responsibility for any analyses, interpretations, or conclusions.

References

1. Belostotsky DA, Sieburth LE (2009) Kill the messenger: mRNA decay and plant development. *Curr Opin Plant Biol* **12**, 96–102.
2. Garneau NL, Wilusz J, Wilusz CJ (2007) The highways and byways of mRNA decay. *Nat Rev Mol Cell Biol* **8**, 113–126.
3. Ramachandran V, Chen X (2008) Small RNA metabolism in Arabidopsis. *Trends Plant Sci* **13**, 368–374.
4. Xie Z, Qi X (2008) Diverse small RNA-directed silencing pathways in plants. *Biochim Biophys Acta* **1779**, 720–724.
5. Gregory B, O'Malley R, Lister R, Urich M, Tonti-Filippini J, et al. (2008) A link between RNA metabolism and silencing affecting arabidopsis development. *Developmental Cell* **14**, 854–866.
6. Addo-Quaye C, Eshoo TW, Bartel DP, Axtell MJ (2008) Endogenous siRNA and miRNA targets identified by sequencing of the Arabidopsis degradome. *Curr Biol* **18**, 758–762.
7. German MA, Pillay M, Jeong DH, Hetawal A, Luo S, et al. (2008) Global identification of microRNA-target RNA pairs by parallel analysis of RNA ends. *Nat Biotechnol* **26**, 941–946.
8. Licatalosi DD, Darnell RB RNA processing and its regulation: global insights into biological networks. *Nat Rev Genet* **11**, 75–87.
9. Coller J, Parker R (2004) Eukaryotic mRNA decapping. *Annu Rev Biochem* **73**, 861–890.
10. Franks TM, Lykke-Andersen J (2008) The control of mRNA decapping and P-body formation. *Mol Cell* **32**, 605–615.
11. Eulalio A, Huntzinger E, Izaurralde E (2008) Getting to the root of miRNA-mediated gene silencing. *Cell* **132**, 9–14.
12. Allen E, Xie Z, Gustafson AM, Carrington JC (2005) microRNA-directed phasing during trans-acting siRNA biogenesis in plants. *Cell* **121**, 207–221.

INDEX

Tamas Dalmay (ed.), *MicroRNAs in Development: Methods and Protocols*, Methods in Molecular Biology, vol. 732, DOI 10.1007/978-1-61779-083-6, © Springer Science+Business Media, LLC 2011

S

T

U

W

X

Z

If you have any concerns about our products,
you can contact us on
ProductSafety@springernature.com

In case Publisher is established outside the EU,
the EU authorized representative is:
Springer Nature Customer Service Center GmbH
Europaplatz 3, 69115 Heidelberg, Germany

Printed by Libri Plureos GmbH
in Hamburg, Germany